C0-AYN-385

中國近代史論著選目

CHINESE COMMUNIST STUDIES OF MODERN CHINESE HISTORY

ALBERT FEUERWERKER

AND

S. CHENG

HARVARD

CHINESE COMMUNIST STUDIES OF MODERN CHINESE HISTORY

by

Albert Feuerwerker

and

S. Cheng

Published by the

East Asian Research Center

Harvard University

Distributed by

Harvard University Press

Cambridge, Mass.

1961

CHINESE ECONOMIC AND POLITICAL STUDIES

This research program at Harvard University seeks to contribute to an understanding of the factors shaping economic and political development in contemporary China. The approach is both historical and interdisciplinary, designed to link contemporary China with its past, particularly the last century of foreign contact, and to pool the skills of history, economics, and political science. These studies are assisted by grants from the Ford Foundation and the Carnegie Corporation.

Special Series

Liang Fang-chung, The Single-Whip Method of Taxation in China

Harold C. Hinton, The Grain Tribute System of China (1845-1911)

Ellsworth C. Carlson, The Kaiping Mines (1877-1912)

Chao Kuo-chün, Agrarian Policies of Mainland China: A Documentary Study (1949-1956)

Edgar Snow, Random Notes on Red China (1936-1945)

Edwin George Beal, Jr., The Origin of Likin (1853-1864)

Chao Kuo-chün, Economic Planning and Organization in Mainland China: A Documentary Study (1949-1957)

John K. Fairbank, Ch'ing Documents: An Introductory Syllabus

Helen Yin and Yi-chang Yin, Economic Statistics of Mainland China (1949-1957)

Wolfgang Franke, The Reform and Abolition of the Traditional Chinese Examination System

Albert Feuerwerker and S. Cheng, Chinese Communist Studies of Modern Chinese History

C. John Stanley, Late Ch'ing Finance: Hu Kuang-yung as an Innovator

CONTENTS

FOREWORD

Great revolutions naturally try to revolutionize the history of the era which preceded them. This is especially true of a Communist-led revolution which believes in using history as a political tool. For American scholars, the historical revolution which the Chinese Communists are attempting has a peculiar interest because it gives "American imperialism" so large a role in determining the course of China's modern history. But quite aside from the Chinese Communists' use or misuse of history, historians everywhere need to know of those scholarly works or documentary collections published in mainland China which contribute to the ongoing stream of historical study.

Toward this end, the most effective preliminary step, in our experience, is an annotated bibliography--not a mere list, but a critical survey--which briefly describes the scope and approach of all major works and brings together in one place the data on items which are of one kind or which concern one topic. It is also most desirable that the critic present in an introduction his own view of the general bibliographical scene he has surveyed. Thereby the user of the volume, in trying to get his own view of this large body of historiography, may also get some view of the critic who has offered to be his guide.

Dr. Feuerwerker, now Associate Professor of History at the University of Michigan at Ann Arbor, undertook this survey in connection with his study of imperialism and its interpretation in late nineteenth-century China, a study which he carried on as a Research Fellow at this Center at Harvard during 1958-59 and 1959-60. He had earlier published a major study in modern economic history, _China's Early Industrialization: Sheng Hsuan-huai (1844-1916) and Mandarin Enterprise_ (Harvard East Asian Studies, No. 1, Harvard University Press, 1958), which raised in concrete form the whole unsettled question of the nature of Chinese society in the late Ch'ing period and the nature of the Western influence upon it. This required an examination of recent Chinese publications on this controversial subject and led Dr. Feuerewerker into a survey of the Chinese

Communists' necessary reinterpretation of China's history. On this latter theme he has published to date two articles, "From 'Feudalism' to 'Capitalism' in Recent Historical Writing from Mainland China," Journal of Asian Studies, 18:107-116 (Nov. 1958), and "China's History in Marxian Dress," American Historical Review, 66:323-353 (Jan. 1961).

Miss Cheng is a researcher in the East Asian Research Center, as well as an instructor in Chinese language. Without her careful perusal, summarizing, and collation of the 500-odd volumes covered in this survey, it would not have been possible. Thus this volume is a product of collaboration carried on at this Center and the Chinese-Japanese Library at Harvard, mainly during 1959-60.

John K. Fairbank

East Asian Research Center
16 Dunster Street
Cambridge, Massachusetts
June 1961

INTRODUCTION: CHINESE COMMUNIST HISTORIOGRAPHY

This volume describes nearly 500 books published in the People's Republic of China between 1949 and 1959, which constitute the main body of Chinese Communist research on modern Chinese history. Our aim in making this reconnaissance has been to provide students of modern China with an introduction to the vast outpouring of historical literature which has occurred in the first decade of the People's Republic of China. By "modern China" we mean particularly the nineteenth and twentieth centuries, although when the nature of the materials has made it advisable, we have pursued some topics into earlier periods, and we have also included in some sections examples of the way in which premodern history is now being treated in Communist China.

Some 2,032 titles, many of them multi-volume works, are listed under "History" in the _Ch'üan-kuo tsung shu-mu_ (National bibliography of China) through December 1958, and undoubtedly other historical works are included elsewhere in its rather cumbersome classification scheme. The vast majority, some 1,674 of these 2,032 works, were published in 1954 and later. Among them is a generous proportion of works intended for the very elementary reader, usually published by the "Youth Publishing House" (Chung-kuo ch'ing-nien ch'u-pan-she), which we have in general excluded from this volume. The more serious works include histories of countries other than China; and in the case of China they encompass the ancient and medieval periods as well as the modern. For example, of the monographic works on relatively restricted subjects published through 1958, 187 titles are concerned with ancient and medieval Chinese history as against 170 titles for modern and contemporary Chinese history. Consequently, although we have not listed and analyzed every important publication on modern history that has appeared in the past ten years, we are confident that our selection is a comparatively comprehensive one, and that the comments that we have made in our introductory notes, on the basis of our selection, would not be significantly changed if we were able to include additional titles.

Our selection has been based, first of all, on the titles available to us. Full access to published materials from Communist China in any field has still not been achieved by any library in the United States. It is for this same reason, inaccessibility, that we have not attempted to survey the flood of historical articles, both popular and scholarly, that have been published in dozens of mainland Chinese journals and newspapers, in the past decade.

Our survey stops with 1959, although two titles published in 1960 have been included. The end of 1959 happens to have coincided with a marked dwindling in the number of historical works from Communist China that have come to our attention, either in the acquisitions of American research libraries or in booksellers' catalogues from Peking, Hong Kong, and Tokyo. What is not clear, at the time of this writing, is whether this shrinkage reflects an actual decrease in the number of historical works being published in the People's Republic of China, or whether it is the result of newly-imposed restrictions on the exportation of such works, similar to those that have been in effect for nearly all except propaganda periodicals and some publications pertaining to the arts since the end of 1959.

The official explanation from Peking for the ending of periodical exports was that a paper shortage had made it necessary for publishers to reduce the size of their editions and thus the copies available for sale outside Communist China. Judging from the extremely poor quality of the paper on which many of the books and periodicals published in the latter part of 1959 were printed, this may in fact have been a legitimate, if only a partial, explanation.[1] If so, it reflects the economic dislocations that have accompanied the "great leap forward"

---

1. On the severe problems of the paper industry in Communist China, see China News Analysis, No. 322 (Hong Kong, May 6, 1960).

in economic development. While we have no firm evidence on which to base this further speculation, the decline in the export, and perhaps in the publication, of historical periodicals and books might also be a sign of the relatively low priority given to the social sciences and the humanities, as opposed to the physical and biological sciences, in the dramatic and costly effort to industrialize which is Communist China's chief preoccupation today. (We have received a circular from Guozi Shudian, dated June 24, 1960, stating that the leading Peking historical journal Li-shih yen-chiu [Historical studies] was again available for export. However, no copies later than October 1959 have yet been received). Foreign students of modern Chinese history, therefore, face the problem that historical writing of the kind and quantity indicated by the present volume probably continues to go on in Communist China while the published results of this mainland historiography are increasingly difficult to procure. Whether this is a temporary condition or one that we shall have to live with for a considerable period, it is as yet impossible to say.

The environment in which the writing and teaching of history in the People's Republic of China has taken place, notwithstanding certain common features, differs in several respects from that of the Soviet Union. In Russia, "it was almost a decade after the Revolution before the Bolsheviks were in a position from the point of view of staff and organization to exert a decisive influence on the historical profession";[2] in post-1949 China, in contrast, history has from the beginning been as much under party control as any other feature of the intellectual scene. But this control has not taken the form of such a person as Pokrovsky, who almost single-handedly acted as dictator over the Soviet historical profession until his death in 1932 and his posthumous disgrace. Nor have historical disputations in Communist China

---

2. C. E. Black, ed., Rewriting Russian History: Soviet Interpretations of Russia's Past (New York: Praeger, 1956), p. 7.

been marked by such personal vilifications, accusations, and counter-accusations as were endemic in the Soviet Union after the death of Pokrovsky. Only during the "anti-rightist" campaign of 1957 were a number of historians severely attacked, and even this was only a minor sideshow to the main spectacle. In recent months, as we shall see below, Professor Shang Yüeh has been singled out and pilloried as a "modern revisionist," but this too seems to be a one-man affair, and the historical profession as a group has not been similarly accused.

In part this relative calm may be explained by the fact that a significant number of prominent historians were already fully committed to the Communist cause prior to 1949. We might mention Wu Han, a Ming specialist, who was active politically during the war and in 1948 left his post at Tsinghua University to join the Communist forces in Manchuria. Others include, of course, Kuo Mo-jo, now president of the Chinese Academy of Sciences; the archaeologist Yin Ta; Fan Wen-lan, director of the Third Office (Modern History) of the Institute of Historical Research; Chien Po-tsan, Hou Wai-lu, Lü Chen-yü, and Pai Shou-i. And even among those historians who were not committed politically, Marxist modes of historical thinking had made considerable inroads during the 1920's and 1930's.

Within this favorable context the Communist regime has sponsored a program of historical revisionism which seeks to remold all of the past on the Marxist-Leninist-Maoist model. In China, where Communism is combined not only with virulent nationalism, but also with the world's longest continuous historical tradition, those who would seek to perpetuate their rule must be acutely aware that however much their goal may be to influence men's minds with respect to the importunate present, realization of that goal is inexorably dependent on the way in which these minds view the past. Mao Tse-tung's "Mandate of Heaven" requires no less justification than that of the Chou conquerors of the earliest Chinese state, the Shang, in the twelfth century B.C.

The guide lines for this task of rewriting the record of China's past were laid down by Kuo Mo-jo in July 1951 in an address to the founding meeting of the Chinese Historical Society: the "old idealistic" view of history was to be replaced by a "materialistic" view. Collective research was to replace individual studies. "Ivory tower" studies, which did not "serve the people," were to be condemned. "Adoration of the past and contempt for the present" were to give way to a greater appreciation of the modern period. An end to "Han chauvinism" and increased attention to the histories of the non-Chinese minority peoples was urged. And, finally, "emphasis on European and American history" was to cede place to the study of Asian history.

The works that have appeared since 1949 seem to hew quite closely to the line set forth by the spokesmen for the historical profession. It is rare to find a volume that does not somewhere between its covers make a bow to the historical theories of Marx and Engels and, increasingly, to the "theoretical writings" of Mao Tse-tung. Multi-volume collections of source materials, often the result of the efforts of committees of compilers, occupy a prominent place in the output of historical works. The writing of modern history, for reasons that we shall note below, was perhaps slower in getting started than the leaders of the Communist Party would have liked, but since the enunciation of the *hou-chin po-ku* ("emphasize the present and de-emphasize the past") line in 1958, special attention has been given to the modern period. A new genre of histories of factories and of communes, theoretically written with the cooperation of the workers and peasants concerned, has recently appeared. The statistics available are inconclusive, but presumably more attention is now being paid to the history of the Chinese Muslims, the Miao, the Mongols, and other minority nationalities. And as before, there is little research on the national histories of countries not in the "socialist camp." (Surprisingly, notwithstanding the abundance of Marxist historical writing in Japan, almost none of it finds its way into Chinese translation.)

Historical research in Communist China is conducted by the

Institute of Historical Research of the Chinese Academy of Sciences, by the several universities, and by ad hoc extra-academic bodies. In addition, the Communist Party itself carries on a certain amount of historical activity, and government archival offices, national and local, from time to time publish collections of historical source materials.

The Institute of Historical Research is part of the Department of Philosophy and Social Science of the Academy of Sciences. Kuo Mo-jo, who is president of the Academy, is simultaneously director of this department. The Institute of Historical Research consists of three offices: the First Office (Ancient History) is also headed by Kuo Mo-jo; the director of the Second Office (Medieval History) is Ch'en Yuan, specialist on T'ang history and the history of Buddhism; the Third Office (Modern History) is directed by Fan Wen-lan, a scholar who began his career as a student of the Chinese classics, but later turned to modern history. Research on economic history is also carried on at the Peking Institute of Economics of the Academy of Sciences, whose deputy director, Yen Chung-p'ing, is one of the most capable scholars in the country. The work of these research organs is carried on both by a permanent research staff and by leading faculty members of universities throughout China who are associated with the several institutes.

Nearly every university and college has a department of history; in some cases there are two departments, one for Chinese history and one for foreign history. These departments, the members of which are of course engaged in both teaching and research, are under the general supervision of the Ministry of Higher Education. While the curricula of the several schools are not uniform, they tend to be quite similar, since many courses are taught from "pedagogical outlines" prepared by conferences of historians which are sponsored regularly by the ministry. In general, the undergraduate history curriculum--there is very little graduate study--extends through four academic years. The ideological content of instruction and research in the field of history, as in other academic areas, is under the constant scrutiny of the Communist Party units in the universities and colleges. We have not been able to determine how many active historians there are at present in the People's Republic of China. It was estimated by the editor of Li-shih yen-chiu that in 1957 there were 1,400 "teachers of contemporary history and history of the Revolution in institutions of higher learning in all parts of the country." This figure refers only to those who are primarily concerned with the period after 1919; it is unlikely that the specialists in any other single period outnumber these modern historians.

A closer look at the activities of a typical department of history, at a medium-rank institution, throws considerable light on the state of history and the historical profession.[3] In November 1958 the department of history at the East China Normal College reported that in the previous three months, as part of its contribution to the "great leap forward," the department had completed 167 "research tasks." The report claimed that the articles and lectures written by the faculty in this period totaled 5,500,000 words, double the total production of this group from the time the college was founded in 1952 until mid-1958. These 167 "research tasks"--we might note here that "research" is being used very loosely--could be divided into five categories. The first consisted of the preparation of seven groups of lecture materials, eleven course outlines, and three study guides; this completed the task of preparing the department's teaching materials which had been incomplete during the first seven years. The second category consisted of writings related to the "thought of Mao Tse-tung," for example "Mao Tse-tung on International Relations," "Deciding the Periodization of Ancient History on the Basis of Mao Tse-tung's Writings." These works, the report stated, "begin to employ the thought of Mao Tse-tung to solve the problems of teaching history and scientific research." In the third group was included the compilation of essays criticizing "the bourgeois historical viewpoint." A public meeting with this object also was held on October 8. The fourth category of research tasks consisted of writing factory

---

3. Li-shih yen-chiu, No. 11: 71-73 (1958); similar reports appear in nearly every issue of this journal.

histories and histories of agricultural producers' cooperatives (the predecessors of the communes). These writings, it was claimed, served to "combine historical science with the reality of life...and with workers' production." This experience "opened a new path for our department's scientific research." The fifth category consisted of monographic writings which stressed contemporary history, as opposed to the alleged emphasis on the premodern period in the past. Many of these papers served an immediate propaganda purpose, for example, an article entitled "Criticism of the New Eisenhower Doctrine."

Why, the report continued, were we able in three months to exceed what we had accomplished in the previous seven or eight years? The answer was because we "followed the mass line under the Party's leadership" and opposed the line that "only specialists can carry out scientific research." The younger teachers on the staff and the students participated fully in the department's research activities. "With many workers, many hands and feet, the quantity of scientific research is correspondingly large, and it is produced that much more quickly; with a large number of persons on the job, and much discussion, the quality of the scientific research is correspondingly high, and its products that much more excellent."

The report concluded by outlining the department's research plan for the following year, through October 1959. Four kinds of activities were projected. First, to continue the preparation of teaching materials, revising those at hand and preparing new outlines and lectures. Second, on the basis of a study of the writings of Mao Tse-tung, to write nineteen critical papers on the bourgeois historical viewpoint, for example, a criticism of T. F. Tsiang's Chung-kuo chin-tai shih (History of Modern China; this is an illuminating personal interpretation of Chinese history since the Opium War, written in 1938 by a leading scholar in the field of China's diplomatic relations who is now the Chinese Nationalist ambassador to the United Nations). Third, to write ten articles in honor of the tenth anniversary of the establishment of the People's Republic of China. Finally, to prepare twenty-eight monographs on modern history, designed to expose imperialist malfeasances and bearing such titles as "French Imperialist Aggression in Shanghai," and "Investigation of American Imperialist Use of Missionaries in Shanghai to Carry Out Aggression in China."

There is, of course, as the contents of this volume will indicate, a considerably better brand of history being written in Communist China than that which this department of history will produce. The tendencies that are so manifest at this institution, however, are increasingly evident in others as well and threaten to engulf even the Institute of Historical Research and the better universities in Peking and elsewhere.

These tendencies are, first, the "mass line," an extension into the field of scholarship of the Communist effort to mobilize the masses in the execution of their program of the "great leap forward" to a modern, industrialized society. As in other fields, given the relatively small number of college graduates in China, the number of "expert" historians is relatively small, and moreover these specialists for the most part received their training under the "old regime" in China or in Europe and America. The "mass line" encourages the relatively ill-trained new crop of teachers, whose recruitment has been made necessary by the rapid expansion of higher education since 1949, and the students--both of which groups have received their education entirely in the Marxist-Maoist milieu--to criticize their "bourgeois" elders and to make up in quantity what they lack in quality by producing vast amounts of popular or semi-popular propaganda writing.

Their scholarly production is to be guided above all by the "thought of Mao Tse-tung," the study of which is continually enjoined on these historians. Almost nothing can be written in the field of modern history which does not take as its starting point a quotation from Mao's writings. This is the second characteristic of current Chinese Communist historiography which is exemplified by the East China Normal College report. There is no better way to deal with an academic opponent than to accuse him of contradicting some dictum from Chairman Mao.

A third trend, closely related to the "mass line," is the writing of factory and commune histories by the staff and students of the several history departments, theoretically with the active cooperation of the workers and peasants. Teachers and students, as well as government and party functionaries, who are sent into the countryside for shorter or longer periods of farm labor alongside the masses, are encouraged to join with the commune members and collectively write histories of the units to which they are attached. Since the autumn of 1958, it is claimed, "tens of thousands of commune histories have been published in various papers and journals, and some have appeared in book form."[4] In a similar manner, factory workers with the assistance of students and other intellectuals are said to have compiled a large number of factory histories in the past two years. Unfortunately, we have seen none of these publications, probably both because they are very ephemeral literature and because they have not been exported. While a certain amount of new material may have been turned up by these efforts, it is likely that the gap between these factory and commune "histories" and anything that might legitimately be called "history" in other countries is rather large. Their purpose is first and foremost propaganda, "to educate and unite the workers as well as to promote production....In the past, even the workers who took part in...struggles weren't entirely clear about their significance. It was only after they began to write down the full story of their experiences that in discussions and informal talks they began to understand the past, and the present, with a clearer insight, see better the course of their own growth and the meaning of working-class solidarity."[5]

---

4. U.S. Consulate General, Hong Kong, Survey of the China Mainland Press, No. 2172:13 (Jan. 8, 1960).

5. Chung Ho, "Workers Write Factory Histories," Peking Review, p. 14 (May 19, 1959).

The "mass line" is a special characteristic of the third source from which historical writing emanates in Communist China, the *ad hoc* groups referred to earlier. Commune histories "written" by local peasants under the leadership of the Communist Party secretary and the functionaries of the commune blend imperceptibly into those undertaken at the initiative of the universities and colleges which we have just discussed. In other cases, local persons, again under Party leadership, may undertake to collect from elderly persons the stories they heard in their youth about the Taiping Rebellion and similar movements; or participants in the 1911 revolution may be encouraged to write down what they remember of the events in which they took part. Several examples of such compilations are included in the text of this volume, and as the readers will readily see, they do not lend much credence to the belief that the "mass line" will overturn the older forms of historical research and composition.

In the upper echelons of the Communist state, execution of the "mass historical line" has taken the form of the establishment in July 1959 of a Written Historical Materials Research Committee of the National Committee of the Chinese People's Political Consultative Conference. This committee has proceeded to collect documents, diaries, and reminiscences from older members of the Political Consultative Conference and from other interested persons. In April 1960 it was claimed that 500,000 words of original source materials and 2,400,000 words of historical materials written by committee members and other persons involved in the project had been collected and edited. These documents covered the period from the end of the Ch'ing dynasty through 1949. For the most part, they seem to be reminiscences written considerably after the fact, albeit often by firsthand observers. By April 1960, four volumes compiled from these materials had been published in a draft form and distributed to members of the National Committee of the Political Consultative Conference. These publications apparently are not yet on general sale; or at least, they cannot be obtained outside of Communist China.[6]

In reading recent reports from several university history departments about their participation in the "great leap forward," we have been struck by a repeated emphasis on the utility of history in the political struggle, the tone of which seems to reflect a fear that Chinese Communist society threatens the historians (and all intellectuals) with the accusation that they are of little use in the building of socialism. The following, by the Department of History at Nankai University in Tientsin, is an example:

> Formerly, when the bourgeois viewpoint in history was rampant, and the past was stressed at the expense of the present, the students would say, "Studying history is of no use." Now having actually participated in historical scientific research, and understanding why and how to study history, they say, "The more we study history, the bigger kick we get out of it". . . .Our faculty and students have all come deeply to realize that the open field for historical science is not in heaps of old paper or in ivory towers, but in the factories and villages, in the midst of the sharp struggles of the masses. Historical science must be in the service of the political struggles and in the service of production. Historical scientific workers must heed the word of the Party and combine with the workers and peasants, combine with reality. Only then can they demonstrate their utility in the socialist revolution and in the construction of socialist society.[7]

Indeed, the mainland historians have not succeeded, in spite of their prolixity, in providing the Communist regime with a neatly packaged past that will call forth unqualified intellectual and emotional

---

6. Jen-min jih-pao (People's daily), p. 19 (Apr. 11, 1960).
7. Li-shih yen-chiu, No. 10: 73-75 (1958).

commitment from their readers. After ten years of ideological remolding, continuous indoctrination in Marxism-Maoism, criticism and self-criticism, and all the other concomitants of political control of intellectual activity, there are still too many loose ends. The deliberate creation of a new, popular, Marxist tradition has apparently aggravated rather than ameliorated the problem of finding meaning in China's past--a quest on which every modern Chinese intellectual, including the Communist historians, must inevitably set forth. It may well be, then, that in these circumstances the historians of Communist China welcomed the new "line" that was proclaimed in the pages of *Li-shih yen-chiu* in May 1958.

That number of the most important historical journal in the People's Republic of China led off with several heavyweight editorials by Kuo Mo-jo, Fan Wen-lan, Ch'en Yuan, Hou Wai-lu, Lü Chen-yü, and Liu Ta-nien, which called upon their colleagues to "emphasize the present and de-emphasize the past." This slogan had a double edge. On the one hand, it was meant to be taken literally, as a guide to the division of the available historical manpower and other facilities, which as in other fields were in short supply. But it also indicated a tightening of the political screws so far as the historians were concerned. If the historians had not done their assigned jobs well, it was because they were still ideologically backward, and a large measure of blame for that backwardness was attributable to their isolation from the great struggles of the masses of workers and peasants to build a socialist society. They had failed to solve the problems of the past because they did not adequately comprehend the present.

Liu Ta-nien, an editor of *Li-shih yen-chiu* and author of a study of American "imperialist aggression" against China, stated it this way: "History after 'May Fourth' [1919] is closely related to everday life.... The object of the study of this phase of our history is to help people to draw a lesson from their activities of yesterday so that they may take part in the practice of today's socialist construction." Liu's "socialist construction" was also meant to embrace the reinterpretation of the past by the historians in order to construct a meaningful heritage: "Marxism tells us that if we are to understand ancient China scientifically, we must emphasize the study of modern and contemporary China." The historical classics of the past and present, Chinese and foreign, he continued, have always reflected certain political and economic systems and have directly or indirectly served such systems. "The Spring and Autumn Annals (Ch'un-ch'iu) of Confucius has been revered as a 'classic,' and therefore it is a typical learned work. But it has been said that Confucius wrote the Spring and Autumn Annals and 'struck fear in the hearts of rebellious ministers and insolent sons.' It is clear that its contents are not 'independent' and not aloof from politics." The burden of Liu's rather skillful reference to Confucius--and also, interestingly enough, to H. B. Morse's International Relations of the Chinese Empire as an example of a classic written from the "bourgeois" viewpoint--was to urge upon the Chinese historians an increased "politicalization" of their work in order "thoroughly to expose the hypocritical viewpoint of the bourgeoisie and establish the Marxist viewpoint instead." "We are willing to do without bourgeois learning; we need only Marxist learning," he concluded.

The hou-chin po-ku line evidently represented a response to the inadequate performance of the historians in providing a new and meaningful past for the People's Republic. From the Party's point of view this was attributable to the persistence of "bourgeois" habits of thought and work. It was also a reflection of the increased general suspicion of intellectuals growing out of the disconcerting "One Hundred Flowers" episode of 1957. In addition, the Communist leaders were concerned over the particularly poor results that the study of contemporary history had so far yielded, a sterility that is reflected in sec. 3.3 of this bibliography. Liu Ta-nien complained that the three leading historical periodicals had published very little in the field of contemporary history. Only 25 per cent of the articles in Li-shih chiao-hsueh (The teaching of history) dealt with the period 1919 and after; the figure for Li-shih yen-chiu was 10.8 per cent; and Wen shih che (Literature, history, and philosophy) had printed no material at all on this period. And even the articles that did appear in print were for the most part source materials and reminiscences; very few were scholarly, scientific pieces of research. If for no other reason, national pride alone required that these recent decades be redeemed from the verdict of "stagnant," "backward," and "decadent" by which both Western and Chinese scholars had hitherto described them. Failure to achieve this redemption, moreover, might cast some shadow on the value of the more distant past; past and present were inextricably related.

It is in response to the slogan "emphasize the present and de-emphasize the past," then, that the mainland historians since 1958 have turned increasingly to writing factory and commune histories and semi-popular propaganda pieces about current events--perhaps even with a sigh of relief. Acceptance of the new line, in spite of the new demands it makes on them, may be seen by the historians as a chance to protect themselves. By plunging into the current "struggles" they can demonstrate that they too have a place and a function in the New China.

It is not likely that the crisis of meaninglessness which Chinese Communist historiography faces can be averted solely by "emphasizing the present and de-emphasizing the past." The reluctance of the problems which the historians have set for themselves--with regard to periodization, peasant revolts, the formation of the Han nation, the role of imperialism, and the like--to yield to solution cannot be ascribed to inferior Marxist tailoring. Quite the contrary. In the specific case of contemporary history (post-1919), and in modern history (1840-1919) as well, how can the historians produce anything other than sterile periodization exercises? Liu Ta-nien admits as much:

> Some comrades think that...the principle of "letting one hundred schools of thought contend" is difficult to carry out [in the study of modern and contemporary history] owing to its close relationship with real life. Those who have new interpretations are afraid of being criticized for revisionism, while those who have no new ideas fear that they will be criticized

for dogmatism. This is a difficult subject where it is easy to run into trouble; and the slogan "emphasize the present" may become "fear the present."

But he immediately brushes the problem aside:

> Contemporary history should not evoke any fear because of its close link with present-day life which makes it practitioners more liable to criticism. That should spur us on to more and more intensive research. The phrase "let one hundred schools of thought contend" (pai-chia cheng-ming) has two meanings: it means free research and also free criticism. If people are free only to make public their views but not to criticize erroneous thinking and erroneous styles of work, then there would be only "crying" (ming) but no "contention" (cheng). But we must have both crying and contention at the same time. There are those who fear that in the research and discussion of contemporary history, academic questions might easily be confused with political questions. Of course the two are related, but they are also distinguishable from each other. In order to distinguish them, and create an environment for free research, academic circles must be able to conduct academic criticism correctly.[8]

Who, in the last analysis, would decide when academic criticism was being conducted "correctly," requires no comment. That the comrades who expressed fear about putting forth new ideas in the writing of modern and contemporary history were not unjustified in their apprehensions is evident from recent developments in the case of Professor Shang Yüeh. Shang (see sec. 4.3 below) has been the leading proponent of the view that the beginnings of capitalism in China should be dated at about the sixteenth century. This position was several times attacked by other

---

8. Li-shih yen-chiu, No. 5:11 (1958).

historians, but for the most part the exchanges in the beginning remained at a scholarly level. It gradually became clear, however (as we explain in the introduction to sec. 4.3) that Shang's views were unacceptable to the Party leadership, because they might tend to minimize the role played by imperialism in transforming China into a semi-colonial, semi-feudal state, and because they might conceivably lead to doubts about the historical necessity of the revolution led by the Chinese Communist Party. In the latter part of 1958 and in 1959 Shang was attacked with increasing severity by more orthodox Party-line historians.[9]

In defending his right to "contend," Shang Yüeh adopted an ingenious stand. Marx, Engels, Lenin, and Mao, when they began to write, he stated, expressed views that clearly were not acceptable to the majority, but these new views were correct ones. No layman, of course, has the right to suggest new interpretations in a field where he is not a master, but the specialist in a particular field of knowledge--for example, an historian who has studied carefully a specific historical question--may arrive at knowledge and insights that in the beginning are possessed by him alone. While these views may not be in agreement with the general opinion, the historian who has new ideas about periodization and the origins of capitalism in China should express them even though they may differ from the "traditional doctrine."[10]

The assertion that the "traditional doctrine" (the "tradition" that Shang referred to was, as everyone knew, the "thought of Mao Tse-tung") could be challenged was sufficient to call forth a torrent of abuse. "In his writings on the Chinese revolution, comrade Mao Tse-tung has made many penetrating analyses of Chinese history. These analyses are objective truths which have been proved in the practice of the Chinese revolution. But Shang Yüeh has slandered them and opposed them as merely 'an old and traditional system of historical analysis,' and has set up

---

9. See, for example, Li-shih yen-chiu, No. 3: 1-11 (1959).

10. Wen-hui pao (Shanghai, Nov. 1, 1959); summarized in China News Analysis, No. 326:7 (June 3, 1960).

his own system of private science which is unrelated to the facts and is anti-Marxist....What Shang Yüeh calls his 'new system' is, however, only the remnants and rotten pieces of the capitalist view of history which he has picked up in a garbage can."[11] In similar tones Shang was attacked as an arch "modern revisionist," who had attempted to revive the capitalist view of history under the cover of "Let the hundred schools of thought contend."

So far as we have been able to ascertain, the pommeling given to Shang has not been extended to any other historians who might dare to hold similar views. It is a warning and an object lesson, however, that the study of modern and contemporary history may indeed "evoke fear because of its close link with present-day life which makes it practitioners more liable to criticism." The works analyzed in the present volume, which leave the study and writing of history in the People's Republic at the end of the first decade of the Communist regime, were produced in this atmosphere of rigid orthodoxy and underlying fear.

For the great majority of the titles in this bibliography we have written commentaries which attempt to indicate the content or kinds of materials contained in each volume and often to assess their value to the researcher. A smaller number of books have been listed by title only. These are usually works of secondary importance; in a few cases they are items which came to our attention too late for inclusion with full commentary in this bibliography. The introductory notes preceding the several sections and the commentaries on adjacent volumes, however, will throw light on those items that have been listed by title only.

At the end of each item there appears in parentheses a number indicating how many copies of the volume cited were published in the edition that we examined. This information is of considerable value in assessing the relative scholarly quality of the titles. Obviously a book published in 100,000 copies is meant for a far wider audience and

---

11. Jen-min jih-pao, p. 7 (June 13, 1960); see also Li-shih yen chiu, No. 4:1-22 (1960) and No. 5:1-48 (1960).

is written in more popular terms than one published in 2,000 copies. In some cases this information, which usually appears on the back of the title page, was not available. These are indicated by a question mark in parentheses.

In the text of this volume, the names of publishers are given only in romanization. A list of publishers with Chinese characters is appended directly before the index. Throughout, we have used the abbreviations CCP (Chinese Communist Party) and KMT (Kuomintang). In view of the prevalence of simplified characters in the publications surveyed, we have made no attempt to substitute the complex forms in our insertions here.

1. GENERAL WORKS

If the titles included in the four subdivisions of Section 1 have any common characteristic, it is that they deal with larger subjects or greater spans of time than those works which fit more or less neatly under the more specific subdivisions of the remaining five sections of this bibliography. The general approach of the Chinese Communist historians in the general survey histories included in sec. 1.1, for example, foreshadows quite clearly the way in which they treat specific historical institutions and events in later sections of this volume. For nearly every subject the reader will want to refer not only to the titles on a specific subject, such as the Taiping Rebellion, but also to the general works in this part. A number of important topics for which the volumes we analyzed were too few to constitute a separate section have been included in sec. 1.1 and sec. 1.4. They might, perhaps, be pointed out here: note especially the items concerned with the history of Taiwan in sec. 1.1; and those dealing with Sino-Soviet relations and Sino-American relations in sec. 1.4.

1.1 GENERAL SURVEYS AND INTERPRETATIONS

The titles discussed under this heading fall into several groups, and in a sense are a microcosm of the entire bibliography. They range from the five doctrinaire survey texts that begin the section to an old-fashioned volume on the historical geography of the city of Wuhan (1.1.30). There is no trace of Marxism-Leninism-Maoism in this last item, and the same may be said of a number of similar publications. But what is important is the degree to which these apolitical academic works are overwhelmed by writings which not only hew consistently to the single line, but also make a conscious show of their unanimity. While old texts and standard sources will surely continue to be reprinted in the Chinese People's Republic, the older academic scholarship (with its vices as well as its virtues) will be an increasing rarity.

Items 1.1.7-9 are collections of historical essays which take up individual problems in Chinese history in the "new" way. Items 1.1.10-13 are examples of the older type of history that still occasionally appear in print, even from the pen of a veteran Communist such as Wu Han. They are followed by several works which treat some of the problems of interpretation that most trouble the mainland historians: the appraisal of individual historical figures (1.1.17-22), the evaluation of peasant uprisings (see also 2.1.20-23, 2.3.3), and the formation of the Han Chinese nation. Next are some examples of current writing on the premodern period of Chinese history which will perhaps give a taste of what is being prepared for those periods not encompassed by this volume. The last four items are concerned with the history of Taiwan.

1.1.1 Chung-kuo li-shih yen-chiu hui 中国历史研究会 (Chinese Historical Society [Fan Wen-lan 范文瀾 et al.]), Chung-kuo t'ung-shih chien-pien 中国通史簡編 (A brief general history of China; Shanghai: Hua-tung jen-min ch'u-pan-she, 1952), 12 + 1072 pp.

This survey, from earliest times down to the Opium War, was one of the earliest Chinese Communist attempts to apply Marxist conceptions, especially the Marxist interpretation of feudalism, to the whole sweep of Chinese history. Chinese society is characterized as feudal from the Western Chou period to the Ch'ing. Originally published as a text in Yenan in 1941, with Fan Wen-lan as the principal author, the volume was reprinted in other parts of Communist-held China and in 1947 in Shanghai. Minor revisions were made in 1948 and the volume republished first in Manchuria in 1949, and then in Shanghai in November 1950 and several times thereafter. The interpretations of Chinese history in this volume have now been somewhat modified by items 1.1.3 and 1.1.4 (38,000)

1.1.2 Shang Yueh 尚鉞 et al., <u>Chung-kuo li-shih kang-yao</u> 中国历史綱要 (An outline of Chinese history; Peking: Jen-min ch'u-pan-she, 1954), 7 + 426 pp., 8 plates.

This is a textbook written originally for use of history students at the Chinese People's University. A substantial amount of material is presented in six well-organized and clearly narrated chapters, beginning with the prehistoric age and ending with the eve of the Opium War. One of the subjects emphasized is the economy, particularly the development of handicraft industry in late Ming which is used to support the thesis that nascent Chinese capitalism would have evolved into a capitalist society even without the intrusion of foreign capitalism. Another subject which gets fairly full treatment is peasant rebellions in the various dynasties. The question of the periodization of ancient history (i.e., the end of slave society and the beginning of feudal society) is bypassed, on the ground that materials are inadequate to support any of the periodization schemes currently advanced. (56,000)

1.1.3 Chien Po-tsan 翦伯贊, Shao Hsun-cheng 邵循正, and Hu Hua 胡華, <u>Chung-kuo li-shih kai-yao</u> 中国历史概要 (A concise history of China; Peking: Jen-min ch'u-pan-she, 1956), 7 + 161 pp.

This outline history is an "official" statement of the current mainland view of China's past, prepared by the authors for the Chinese Historical Association (Chung-kuo shih-hsueh hui 中国史学会) and circulated among other historians for their suggestions before the final draft was written. China's history is divided into three parts: (1) Ancient, from the legendary Hsia dynasty to the Opium War of 1840, and encompassing in turn primitive society, slave society, and feudal society (pp. 1-50).

The only significant forces of progress were the three peasant uprisings at the end of the Ch'in, Sui, and Yuan dynasties, "on the foundations of which arose the three illustrious empires of the Han, T'ang, and Ming." (2) Modern (the period of the "old democratic revolution"), meaning the years 1840-1911, when China became semi-colonial and semi-feudal (pp. 51-83). Here the important events were the Taiping and Boxer movements, and the 1911 revolution, all of which failed because they had no proletarian leadership. (3) Contemporary (the period of the "new democratic revolution and transition from new democracy to socialism"), from 1919 to the present (pp. 84-161). Major stages are the May Fourth Movement, the founding of the CCP, the "First and Second Revolutionary Civil Wars," the war against Japan, the "Third Revolutionary Civil War," and finally, the establishment of the Chinese People's Republic and its "progress on the road to socialism." The relative amount of space given to the three periods--ancient: 4000 years, 50 pages; modern: 79 years, 33 pages; and contemporary: 30-odd years, 78 pages--is a good indicator of the pressure of the present on all Chinese historiography. (26,000)

1.1.4 Lü Chen-yü 吕振羽, Chien-ming Chung-kuo t'ung-shih 簡明中国通史 (A concise history of China; Peking: Jen-min ch'u-pan-she, 1959), 8 + 976 pp.

The third edition, now in one volume, of a textbook by a pioneer Marxist historian which has become a standard work in mainland China. Vol. 1, first published in 1941, and Vol. 2, first published in 1947, went through printings totaling more than 400,000 copies before the author undertook minor revisions in 1951 and 1953. This second edition was first issued in 1955 and appeared in several printings totaling 100,500 copies. The present edition, in contrast, includes considerable substantive revision, in particular in the sections dealing with three particularly sensitive areas in current Chinese historiography. (1) Lü's latest text is much more equivocal on the question of the division point between the period of slave society and that of feudalism in ancient China. He still holds that the Western Chou was a "transitional era," but now places great stress on the "unevenness" in the economic and social development of the several feudal states. (2) In the controversy over the beginnings of capitalism in China he carefully separates himself from the position of Shang Yüeh (see sec. 4.3) and gives less weight than in earlier editions to the degree of development toward capitalism which had occurred prior to the Opium War. (3) The third edition goes somewhat beyond the second (which itself represents a major change from the first) in eliminating overtones of Han chauvinism *vis-à-vis* the non-Chinese minority nationalities (see also 1.1.5). Reference notes to each chapter have been added. (142,500)

1.1.5 Lü Chen-yü 吕振羽, *Chung-kuo min-tsu chien-shih* 中国民族簡史 (A brief history of the peoples of China; Shanghai: San-lien shu-tien, 1950), 30 + 218 pp.

A short and only partially documented study of the peoples or nationalities of China--Han, Manchu, Mongol, Tibetan, and a dozen or so other tribal groups--with reference to their origins, special characteristics, historical development, and the like. The author believes that intermarriage and cultural interchange have long ago eliminated the racial boundaries of these groups. He attacks in particular writers who have attributed racial and cultural superiority exclusively to the Han. (10,000)

1.1.6 Shu Shih-cheng 束世澂, *Chung-kuo t'ung-shih ts'an-k'ao tzu-liao hsuan-chi* 中国通史参考資料選輯 (Selected reference materials on the general history of China), 8 vols. (Shanghai: Hsin chih-shih ch'u-pan-she, 1955), Vol. 1, 160 pp., illus.

The first volume of this eight-volume publication is on the prehistoric era and contains some forty short selections from the writings of archaeologists on the early civilization of different regions in China in several periods including the Bronze Age. The final portion of Vol. 1 consists of quotations from the classics (e.g. Confucius, *Tso-chuan* 左傳, *Shih-chi* 史記) which refer to ancient legends about the sources of Chinese civilization. The remaining seven volumes deal with periods from the Shang through the Ch'ing dynasty before 1840, but unfortunately were not available for our examination. (28,000)

1.1.7 Lü Chen-yü 呂振羽, *Chung-kuo she-hui shih chu wen-t'i* 中国社会史諸問題 (Some questions concerning the history of Chinese society; Shanghai: Hua-tung jen-min ch'u-pan-she, 1954), 6 + 210 pp.

An earlier version of this book was published in Chungking in 1940. Changes in the present edition are, according to the author, largely editorial (e.g., terms and phrases that were "doctored up" for the benefit of the wartime Chungking regime are now replaced by words more consistent with the author's viewpoint and argument). The major subjects discussed are the "Asiatic mode of production," the "stagnation" of Chinese society, the slave system in Chinese history, and the creation of a new national culture and the question of cultural inheritance. The appendix, "An outline for the study of Chinese history," gives in capsule form a good idea of the chief preoccupations of Chinese Communist historiography. (24,000)

1.1.8 Chien Po-tsan 翦伯贊, Li-shih wen-t'i lun ts'ung 历史問題論叢 (Collected essays on historical questions; Peking; San-lien shu-tien, 1956), 197 pp.

Seven of these eight essays were originally published between 1951 and 1954. Two essays deal with the "proper way" to study Chinese history, viz--by making use of dialectical materialism. The subjects of the other articles are: the evaluation of historical personages, archaeological discoveries and historical studies, feudal society in ancient China, peasant wars in ancient China (the last two were written as commentaries on the chapter on feudal society in Mao Tse-tung's The Chinese Revolution and the Chinese Communist Party), and slavery in the two Han dynasties. The eighth essay, the longest in the collection (pp. 117-193), entitled "On the nature of the Chinese economy in the first half of the eighteenth century," is Chien Po-tsan's interpretation of the economic background in Ts'ao Hsueh-ch'in's novel Hung-lou meng (Dream of the Red Chamber). It was read before the 1955 Conference of Junior Sinologues in Leiden. (11,000)

1.1.9 Chang Shun-hui 張舜徽, Chung-kuo-shih lun-wen chi 中国史論文集 (Collected essays on Chinese history; Wuhan: Hu-pei jen-min ch'u-pan-she, 1956, 5 + 201 pp.

The most useful of the ten articles in this volume is a partial listing by subject (with chüan number) of the contents of the Huang-Ming ching-shih wen-pien 皇明經世文篇 (Writings on statecraft of the Ming dynasty), com. Ch'en Tzu-lung, pp. 131-139. Other items discuss these subjects: (1) an historical view of the simplification of Chinese characters; (2) materials for the study of ancient Chinese history; (3) important points for beginning students of oracle bones and bronze inscriptions (this contains criticisms of some of Kuo Mo-jo's work); (4) the use of historical sources; (5) love and hate as expressed in popular tales among the laboring people (ending with the Ming period); (6) peasant uprisings during the Sung dynasty; (7) the contribution of the Sung scholars; (8) the work of two archaeologists, Lo Chen-yü 羅振玉 and Wang Kuo-wei 王國維. Ten letters from the author to his friends, containing his views on topics related to the study of Chinese history, are included in an appendix. (16,000)

1.1.10 Chou Ku-ch'eng 周谷成, Chung-kuo t'ung-shih 中国通史 (A general history of China), rev. ed., 2 vols. (Shanghai: Hsin chih-shih ch'u-pan-she, Vol. 1, 1955; Vol. 2, 1956), 7 + 441 + 7 + 448 pp.

The first edition (1939) of this general survey, from prehistoric times down to about 1930, already used a broad Marxist periodization scheme: formation of feudal society (A.D. 9-960); continuation of feudal society (960-1840); era of incipient capitalism (1840-1920's). The present edition ends with the "Twenty-one Demands" in 1915, but retains basically the same material as the first edition, including frequent quotations from the standard Chinese sources. There are some significant changes, however, in the organization of material--for example, the Taiping Rebellion has now been moved from the end of the feudal period to the beginning of the modern period, since the anti-foreign role of the rebellion is currently being stressed. A striking difference appears in the introductions to the two editions: in the 1939 edition, the author comments that Marx's statement that history is the history of class struggles constitutes but a view of history, not history itself. In the 1955 edition, the author says that history is the "process of [class] struggles." (20,000)

General Surveys and Interpretations

1.1.11 Ch'en Teng-yuan 陳登原, *Kuo-shih chiu wen* 國史舊聞 (Old tales of Chinese history; Peking: San-lien shu-tien, 1958), Vol. 1, 12 + 680 pp.

This work, written in the literary style, was undertaken shortly after 1938 when the author retired from Chekiang University. Vol. 1 deals with 283 miscellaneous topics related to Chinese history in roughly chronological order up through the Sui period. Under each topic are presented quotations taken from standard historical works (e.g. *Shih-chi*, *Han-shu*) and other sources. These are followed by notes by the compiler which comment on the validity, significance, or other aspects of the evidences presented. Topics concern principally the growth of political and social institutions and material culture. (4,000)

1.1.12 Teng Chih-ch'eng 鄧之誠 *Ku-tung so-chi ch'üan-pien* 骨董瑣記全編 (Reading notes on historical subjects; Peking: San-lien shu-tien, 1955); 28 + 656 pp.

Presumably a reprint of the 1933 edition; it was not available for our examination. (?)

1.1.13 Wu Han 吳晗, *Tu shih cha-chi* 讀史劄記 (Reading notes on history; Peking: San-lien shu-tien, 1956), 359 pp.

A collection of eleven scholarly articles originally written and published between 1931 and 1948. The topics are: (1) the system of "captives' households" (*t'ou-hsia* 投下) in the Liao and early Ming periods; (2) the currency system of the Yuan period; (3) Wang Mao-yin 王茂蔭 (1798-1895) and the currency reform of the Hsien-feng period; (4) Yi Man-ju 李滿住 of the *Yi Dynasty Veritable Records* 李朝實錄. The remaining articles are on the Ming period, dealing specifically with: (1) the military system, (2) the paper currency, (3) the academies, (4) the compilation of the Ming shih-lu 明實錄 (Veritable records of the Ming), (5) Manichaeism and the Ming empire, (6) the social background of the novel Chin P'ing Mei 金瓶梅, and (7) the scholar-official Ch'ien Ch'ien-i 錢謙益 (1582-1664). (5,000)

1.1.14 Jung Meng-yuan 榮孟源, Li-shih jen-wu ti p'ing-chia wen-t'i 历史人物的評價問題 (The question of appraising historical personages; Shanghai: Hua-tung jen-min ch'u-pan-she, 1954), 2 + 65 pp.

Though Mr. Jung begins by quoting Ssu-ma Ch'ien, Pan Ku, and Liang Ch'i-ch'ao, to illustrate the importance of evaluating individuals in history, he dismisses them and traditional Chinese historiography as "feudalistic" and "restricted by a capitalistic class bias." With this as his point of departure, he offers a series of ideological exercises on the role of the individual in history which are of interest only because of their application of Marxist theory to relatively unfamiliar Chinese materials by someone who has a good command of traditional historiography. (35,000)

1.1.15 Ch'en Hsü-lu 陳旭麓, Lun li-shih jen-wu p'ing-chia wen-t'i 論历史人物評價問題 (On the question of evaluating historical personages; Shanghai: Hsin chih-shih ch'u-pan-she, 1955), 44 pp. (31,000)

1.1.16 Li Kuang-pi 李光璧, Ch'ien Chün-hua 錢君曄, eds., Chung-kuo li-shih jen-wu lun chi 中国历史人物論集 (Collected essays on Chinese historical personages; Peking: San-lien shu-tien, 1957), 372 pp.

This volume consists of twenty essays which attempt to appraise or reappraise Chinese historical figures between the Ch'in and Ch'ing dynasties. Most of the subjects are well-known statesmen, military, or intellectual leaders (e.g., Wang Mang, Ts'ao Ts'ao, Chu-ko Liang, Ssu-ma Ch'ien); one or two are relatively less known (e.g., Shen K'uo 沈括, 1031-95; T'ang Chen 唐甄, 1630-1704). The editors' ideal for the analysis of historical persons and their historical role is stated to be "the application to historical facts of the Marxist-Leninist position, viewpoint, and method." All the essays do not measure up to this ideal to the same degree. Nevertheless, they employ a good deal of current Chinese Communist terminology, and the conclusions they draw illustrate some of the principal concerns of contemporary Chinese historiography. (20,000)

1.1.17 Yang K'uan 楊寬, Ch'in shih-huang 秦始皇 (Biography of Ch'in shih-huang; Shanghai: Shang-hai jen-min ch'u-pan-she, 1956), 2 + 124 pp. (25,000)

1.1.18 Chang Wei-hua 張維華, Lun Han Wu-ti 論漢武帝 (On Emperor Wu-ti of the Han dynasty; Shanghai: Shang-hai jen-min ch'u-pan-she, 1957), 179 pp. (?)

1.1.19 Wang Chung-lo 王仲犖, Ts'ao Ts'ao 曹操 (Biography of Ts'ao Ts'ao; Shanghai: Shang-hai jen-min ch'u-pan-she, 1956), 2 + 130 pp. (40,000)

1.1.20 Teng Kuang-ming 鄧廣銘, Wang An-shih 王安石 (Biography of Wang An-shih; Peking: San-lien shu-tien, 1953), 107 pp., 1 plate. (?)

1.1.21 Teng Kuang-ming 鄧廣銘, Yueh Fei chuan 岳飛傳 (A biography of Yueh Fei; Peking: San-lien shu-tien, 1955), 4 + 303 pp. (45,000)

1.1.22 Yü Yuan-an 余元盦, Ch'eng-chi-ssu-han chuan 成吉思汗傳 (A biography of Chinggis Khan; Shanghai: Shang-hai jen-min ch'u-pan-she, 1955), 98 pp., 1 map. (?)

1.1.23 Li-shih chiao-hsueh yueh-k'an she 历史教学月刊社 "(History instruction monthly" society), ed., Chung-kuo nung-min ch'i-i lun chi 中国農民起義論集 (Collected essays on Chinese peasant uprisings; Peking: Wu-shih nien-tai ch'u-pan-she, 1954), 3 + 200 pp.

The viewpoint of this volume is summarized in these words of Mao Tse-tung which are quoted in the Foreword: "In Chinese feudal society, only...peasant class struggles, peasant uprisings, and peasant wars furnish the real motive force of history." The editors further state that they attempt to "expose or criticize previous historical writings which distort or insult peasant uprisings, and seek to employ historical facts in order to promote the superb heritage of glorious participation of the fatherland's laboring people." (Thus the term ch'i-i 起義, "righteous uprising," is used consistently for revolts of all types and magnitudes). To these ends, the volume assembles fifteen research articles of varying length which deal with peasant risings from the "Yellow Turbans" of the Han to the hitherto little-known peasant rebellion led by Sung Ching-shih 宋景詩 of the late nineteenth century. But note that, explicitly or implicitly, the ultimate failure of all these rebellions is attributed to the fact that they were peasant uprisings and thus incapable of founding a new socio-economic system. (15,000)

1.1.24 Li Kuang-pi 李光璧, et al., eds., Chung-kuo nung-min ch'i-i lun chi 中国農民起义論集 (Collected essays on Chinese peasant uprisings; Peking: San-lien shu-tien, 1958), 2 + 462 pp.

This collection incorporates fourteen out of the fifteen essays, revised in varying degrees, originally found in a collection published in 1954 (1.1.23), plus twelve new essays. The new material spans a longer period, from the end of the Ch'in dynasty down to the early Republic. The same implicit and loose definition of "uprising" applies in this volume as in the previous one. A short article on an early Northern-Sung rebellion in the earlier collection is replaced in this one by a more fully developed article on the same subject. The additional topics include the uprising of the Uigurs in Sinkiang in the nineteenth century, the White Lotus and Boxer rebellions, and an account of Pai Lang 白朗, the leader of a peasant uprising in North China between 1912 and 1914. (6,000)

1.1.25 Chao Li-sheng 趙儷生 and Kao Chao-i 高昭一, Chung-kuo nung-min chan-cheng shih lun-wen chi 中国农民战争史論文集 (Collected essays on the history of Chinese peasant wars; Shanghai: Hsin chih-shih ch'u-pan-she, 1955), 163 pp., 5 maps.

This volume is a byproduct of a new course on the "History of Chinese Peasant Wars" which was offered at Shantung University in 1953. In its eleven essays the authors cover the period from the Ch'in-Han to the late Ming and rely largely on the "standard histories" for their studies of "peasant wars." The first item rather interestingly asserts that although peasants naturally were the chief participants in peasant revolts, the instigators of revolt "as a rule included handicraft workers, poor city-dwellers, and even small shopkeepers, intellectuals, and middle and small landowners." Typical of the contents is an article on the activities of the remnants of the forces of Li Tzu-ch'eng (pp. 154-163) after the Manchu conquest. (28,000)

1.1.26 Sun Tsu-min 孫祚民, Chung-kuo nung-min chan-cheng wen-t'i t'an-so 中国农民战争问题探索 (An inquiry into problems concerning peasant wars in China; Shanghai: Hsin-chih-shih ch'u-pan-she, 1956), 3 + 95 pp.

Eight essays, previously published in Li-shih yen-chiu and other journals, discuss the proper historical approach to "peasant wars," a term loosely employed to indicate end-of-dynasty revolts and other popular uprisings. What is their "true" social character, as opposed to the traditional view? What is their role in Chinese history? What type of "political power" do peasant rebels establish? What is the relation of peasant wars to national unity, to uprisings of minority races, and to religion? How shall the leaders of peasant uprisings be evaluated? These discussions are heavily buttressed with quotations from Marx, Lenin, and Mao Tse-tung. (18,000)

1.1.27 Liu K'ai-yang 刘開揚, Ch'in-mo nung-min chan-cheng shih-lueh 秦末农民战争史略 (A brief history of peasant wars at the end of the Ch'in dynasty; Peking: Commercial Press, 1959), 98 pp., 2 maps. (10,000)

1.1.28 Ch'i Hsia 漆侠, Sui-mo nung-min ch'i-i 隋末农民起义 (Peasant uprisings at the end of the Sui dynasty; Shanghai: Hua-tung jen-min ch'u-pan-she, 1954), 2 + 120 pp.

This small volume offers a somewhat popularized account of the tyrannical rule of Sui Yang-ti (604-610), the peasant rebellions that finally killed him (610-618), and the ensuing period of civil war until the completion of the T'ang unification (618-624). The principal sources are the Sui shu, T'ang shu, and Tzu-chih t'ung-chien. Although the author notes in a post-script that research on peasant risings in the Sui is rendered difficult by the fragmentary and sometimes conflicting nature of the historical record, this does not prevent him from concluding his book with a Maoist moral about the role of the peasant wars in Chinese history. (64,000)

1.1.29 Han min-tsu hsing-ch'eng wen-t'i t'ao-lun chi 漢民族形成問題討論集 (Collected essays on the problem of the formation of the Han nation), compiled by the editors of Li-shih yen-chiu 历史研究 (Historical studies; Peking: San-lien shu-tien, 1957), 2 + 262 pp.

Like the matter of periodization, the "problem of the formation of the Han nation," the editors point out, is one that is unsolved in Chinese historiography. These eleven articles will not settle the matter definitively. All but one were published previously in historical journals between 1954 and 1956. Pages 228-254 contain in translation the article by Prof. G. V. Efimov of Leningrad University (Voprosy istorii, Oct. 1953) which gave rise to the discussions, and pp. 255-262 summarize the state of the discussions as of late 1955. The argument is largely between those who support Fan Wen-lan's broader interpretation of the Marxist-Stalinist critieria of a "nation" (pp. 1-16) and those who hold to a more strict view. This debate is, however, far removed from Chinese historical reality and in particular from what it is ostensibly considering, the idea of a Chinese "nation" and "nationalism." (9,600)

1.1.30 P'an Hsin-tsao 潘新藻, Wu-han-shih chien-chih yen-ko 武漢市建制沿革 (The successive stages in the founding of the city of Wuhan; Wuhan: Hu-pei jen-min ch'u-pan-she, 1956), 8 + 70 pp., 10 maps.

In tracing the history of the modern city of Wuhan back to the Han dynasty, this volume first investigates the changing course of the Han River (for which it relies heavily on Shui-ching chu 水經注, the "Water Classic" of Northern Wei). Secondly, it discusses the sites of ancient forts and cities in the area, citing local gazetteers and the like. Finally, it examines the historical stages of the city now called Wuhan, referring to Shih-chi, Shang-shu, and the dynastic histories as the principal sources. This section brings the subject briefly down to 1954; but it is apparent that the author's main interest is a traditional type of historical geography. (3,500)

1.1.31 Ch'i Ssu-ho 齊思和, Chung-kuo ho Pai-chan-t'ing ti-kuo ti kuan-hsi 中国和拜占廷帝国的関係 (China's relations with the Byzantine Empire; Shanghai: Shang-hai jen-min ch'u-pan-she, 1956), 37 pp.

The author (a Harvard-trained historian) first examines references to the Byzantine Empire in the Chinese historical record (e.g., the Wei, T'ang, and Sung histories), and correlates them with the works of Western historians (e.g., Gibbon, Lindsay, Diehl). He then discusses the introduction of silkworm culture from China into the Byzantine Empire, and, in the opposite direction, the importation of precious stones, fabrics, and Nestorian Christianity into China. Copious footnotes refer to extensive Chinese and Western sources. (9,000)

1.1.32 Ts'en Chung-mien 岑仲勉, Huang-ho pien-ch'ien shih 黃河變迁史 (A history of the changes of the Yellow River; Peking: Jen-min ch'u-pan-she, 1957), 2 + 786 pp., 10 maps.

A well-documented account of the Yellow River problem and efforts to meet it, mainly in the ancient periods up through the Yuan dynasty (to p. 461). A summary of Ming and Ch'ing efforts (pp. 462-659) is followed by a brief section on recent developments, a bibliography (5 pp.) and a list of place names. (5,000)

1.1.33 Ho Tzu-ch'üan 何茲全, Wei Chin Nan-pei ch'ao shih lueh 魏晉南北朝史略 (A brief history of the Wei, Chin, and Southern and Northern dynasties; Shanghai: Shang-hai jen-min ch'u-pan-she, 1958), 3 + 224 pp.

Derived from the author's lectures at the Peking Normal University, this volume deals in considerable detail with the years ca. 220-580. In addition to tracing the complex political and military conflicts of the Six Dynasties period, the author pays some attention to the dominant economic features (e.g., Ts'ao Ts'ao's land system). In addition, for each of the three epochs into which he divides his narrative (the Three Kingdoms and Western Chin, Eastern Chin, and the Northern and Southern dynasties) there is a separate section summarizing the state of poetry, philosophy, religion, and the like. The sources are for the most part the dynastic histories. There are only a few references to Marx (e.g., in a short section on the "feudalization of the Toba tribe") and in general the text is freer of clichés than are many of its kind, although such analysis as the volume contains is quite orthodox Marxism-Leninism. (14,000)

1.1.34 Han Kuo-p'an 韓国磐, Sui-ch'ao shih lueh 隋朝史略 (A brief history of the Sui dynasty; Shanghai: Hua-tung jen-min ch'u-pan-she, 1954), 4 + 80 pp.

An elementary survey of Sui history, touching on: (1) the founding of the dynasty and the unification of China; (2) the economy, including agriculture and the Grand Canal; (3) external relations, including wars and trade; (4) peasant risings and the fall of the dynasty; and (5) cultural achievements. The style is completely vernacular; even direct quotations are translated. The only footnotes refer to direct quotes. What analysis is included is a rather vulgar Marxism (e.g., the reason why Sui was able to unify China was the greater advancement of agriculture in the Northern dynasties than in the Southern). (28,000)

1.1.35 Ts'en Chung-mien 岑仲勉, Sui T'ang shih 隋唐史 (A history of the Sui and T'ang dynasties; Peking: Kao-teng chiao-yü ch'u-pan-she, 1957), 12 + 681 pp.

Based on Professor Ts'en's pre-1949 lectures at Sun Yat-sen University, Canton, this volume deals first with the Sui dynasty in nineteen fairly short sections which discuss, for example, the reunification of China, centralization, relations with the Turkish tribes of Central Asia, construction projects, the tyranny of Sui Yang-ti, trade, the economy, and the catastrophic military expeditions to Korea. Appended to Part I are two short exercises which try to discuss Sui history in terms of dialectical materialism. Part II, on the T'ang (pp. 91-663), contains sixty-eight sections which discuss in great detail and in a traditional manner political events, administrative and economic institutions, and the intellectual and cultural record. The book is written in the literary style, ostensibly for reasons of economy of words and thereby of paper. A list of reference works appears on pp. 666-681. (12,000)

1.1.36 Yang Chih-chiu 楊志玖, Sui T'ang Wu-tai shih kang-yao 隋唐五代史綱要 (An outline of the history of the Sui, T'ang, and Five Dynasties; Shanghai: Shang-hai jen-min ch'u-pan-she, 1957), 7 + 165 pp. (12,000)

1.1.37 Ch'i Hsia 漆俠, Wang An-shih pien-fa 王安石变法 (Wang An-shih's reform; Shanghai: Shang-hai jen-min ch'u-pan-she, 1959), 3 + 256 pp.

In this book Wang An-shih is reinterpreted in Marxist terms, from which treatment he emerges, like many another reformer, as a "product of class contradictions," a "progressive force" doomed to failure because he was not in line with the peasants and their struggles. Before proceeding with this story, the author devotes his introduction to the criticism of the "fallacious" theories of the "bourgeois" historiographers of the Sung reformer, including Ts'ai Shang-hsiang 蔡上翔, Liang Ch'i-ch'ao 梁啟超, Hu Shih 胡適, and Ch'ien Mu 錢穆. The substantive part of the volume is contained in five chapters dealing with the political and economic institutions of the Sung "feudal" society, the sporadic attempts at reform before Wang An-shih, the content of Wang's proposed reforms, the vicissitudes of the reform program and the opposition, and finally the failure of the reform. An appendix (pp. 237-254) contains some remarks on sources and the text of Wang's "new laws" or "new policies." The text is detailed and well documented, and should be of use despite its foregone conclusions. (7,000)

1.1.38 Li Chih-fu 李稚甫, T'ai-wan jen-min ko-ming tou-cheng chien-shih 台湾人民革命鬥爭簡史 (A concise history of the Taiwan people's revolutionary struggles; Canton: Hua-nan jen-min ch'u-pan-she,

1955), 4 + 196 pp.

This brief account of the resistance of the Taiwanese to foreign invasions since the seventeenth century is written in a heavily propagandist tone and for propaganda purposes: to increase the Chinese people's realization of their "sacred task to liberate Taiwan." Following the assertion that Taiwan was Chinese territory from ancient times, the text centers on these periods: (1) 1662-1840, struggles against the Ch'ing; (2) 1841-95, resistance to Western invasions; (3) 1895-1945, resistance to Japanese rule; (4) 1945-47, persecution by Chiang Kai-shek with the support of the U.S.A., and the "February 28 revolution." (7,200)

1.1.39 Ch'ien Chün-hua 錢君曄 and Yang Ssu-shen 楊思慎, T'ai-wan jen-min tou-cheng chien-shih 台湾人民鬥爭簡史 (A brief history of struggles of the Taiwanese people; Tientsin: T'ien-chin jen-min ch'u-pan-she, 1956), 76 pp., 4 plates, 2 maps.

A popular account of the history of Taiwan which dates its earliest contacts with China in the third century and briefly surveys the development of the island by Chinese settlers under Koxinga in the seventeenth century, and by Governor-general Liu Ming-ch'uan 劉銘傳 in the late nineteenth century. For the years 1657-1947 the narrative centers on the successive "struggles" of the Taiwanese against the Dutch, the Ch'ing, Britain and France, Japan, and American imperialism and Chiang Kai-shek. All this leads up to the conclusion that Taiwan must now be returned to mainland rule. (10,820)

1.1.40 Liu Ta-nien 刘大年, Ting Ming-nan 丁名楠, and Yü Sheng-wu 余繩武, T'ai-wan li-shih kai shu 台湾历史概述

(A general account of the history of Taiwan; Peking: San-lien shu-tien, 1956), 79 pp., 1 map, illus. (19,000)

1.1.41 Wu Chuang-ta 吳壯達, <u>T'ai-wan ti k'ai-fa</u> 台湾的開發 (The development of Taiwan; Peking: K'o-hsueh ch'u-pan-she, 1958), 3 + 82 pp. (2,650)

1.2 ON THE PERIODIZATION OF CHINESE HISTORY

The proper dating of the slave and feudal stages of Chinese history and of the beginnings of capitalism, and the division of modern history into its appropriate parts, have since 1949 been central concerns of Chinese Communist historiography. The goal of the continuous discussions of periodization is to fit Chinese history into a Marxist suit of clothes, with the end product of these procrustean labors to be a Chinese history that they confidently assume can be valued because it was inevitable and because, notwithstanding a long period of stagnation, in a last-minute spurt it has completed the prescribed course well ahead of its competitors.

It is assumed that Chinese society passed from primitive communism through slavery, feudalism, and capitalism to socialism, and that it will soon achieve the goal of complete communism. But when did the era of slavery end in China and when did the feudal era begin? This is the central question in the periodization of ancient history. The pressure to arrive at a final answer to this (and to other questions of periodization as well) probably stems in part from the anxiety of the Communist Party leadership lest any ambiguity as to the early stages of the historical process raise doubts about its completion. Items 1.2.1-6 deal with the question of dividing up ancient history. Apparently relatively less attention has been given to the two millennia of "feudal" society that the mainland historians assert follow the epic of slavery.

This is perhaps indicative of a reluctance to venture forth into uncharted waters. The Marxist classics have had very little to say about the periodization of the feudal epoch; their emphasis has always been on tracing the origins and development of a capitalism that successfully grew out of medieval society and much less on analyzing the antecedent feudal order. Item 1.2.7 is an example of how the feudal period is being treated. The nearer we come to modern times, the greater is the agreement among historians and the more is conformity apparently required. With respect to modern history proper, the consensus on the main outlines and characteristics of its periodization seems nearly complete. Item 1.2.8 contains materials summarizing the discussions of the periodization of modern history. See also sec. 4.3 on the controversy over the origins of capitalism in China.

1.2.1 Hou Wai-lu 侯外廬, Chung-kuo ku-tai she-hui shih 中国古代社会史 (A history of ancient Chinese society; Peking: San-lien shu-tien, 1949), 8 + 337 pp.

Fourteen essays on ancient Chinese society by an historian who has enjoyed a measure of esteem as a pioneer in applying Marxist theory to China. He deals with such topics as the "Asiatic mode of production" (which he sees as a variation of classical slave society), the Chinese "city-state," and reform movements in ancient China. The text was written in 1946 and, since the publication of this edition, has been revised to correct errors and misprints and to adjust the terminology which originally had been phrased to meet the requirements of KMT censorship. The new title is Chung-kuo ku-tai she-hui shih lun 中国古代社会史論 (Essays on ancient Chinese society; Peking: Jen-min ch'u-pan-she, 1955), 388 pp. (8,000)

1.2.2 Chung-kuo ti nu-li-chih yü feng-chien-chih fen-ch'i wen-t'i lun-wen hsuan-chi 中国的奴隸制与封建制分期問題論文选集 (Collected essays on the problem of periodizing China's slave and feudal systems), compiled by the editors of Li-shih yen-chiu 历史研究 (Historical studies; Peking: San-lien shu-tien, 1956), 4 + 508 pp.

The attempts to fit China's ancient history into the Marxian boxes of "slave society" and "feudal society," and have nothing left over--have been going on now for at least thirty years. On the evidence of these twenty-five articles (which first appeared in a variety of mainland periodicals between 1950 and 1955) the problem is no more near definitive solution now than it was in the 1930's. The chief protagonists are Kuo Mo-jo 郭沫若 and Fan Wen-lan 范文瀾, and the issue between them is how to characterize the Western Chou. For Kuo, it is still a slave society; for Fan, China entered the feudal stage with the Chou conquest in the eleventh century B.C. The twenty-one articles that are not by Kuo or Fan harmonize on these two themes in extensive scholastic detail. (5,000)

1.2.3 Chung-kuo ku-shih fen-ch'i wen-t'i lun-ts'ung 中国古史分期問題論丛 (Essays on the periodization of Chinese ancient history), compiled by the editors of the magazine Wen shih che 文史哲 (Literature, history, and philosophy; Peking: Chung-hua shu-chü, 1957), 359 pp.

Fourteen articles on the periodization of ancient history which appeared originally in Wen shih che during 1956 and 1957 are reprinted with minor changes. The first three outline the problems that perplex CCP historians of the pre-Han era, in particular that of when "slave society" came to an end and "feudal society" began, and the nature of early "feudalism." The ten articles that follow are detailed substantive examinations of specific problems concerning the Chou period and especially the critical Western Chou (e.g., the land and tax systems, the nature of slavery.) Throughout there is considerable reference to Soviet writings on this period and these problems, and the final item reprints an exchange of letters between Professor T'ung Shu-yeh 童書業 of Shantung University and a Soviet colleague which reflects disagreement on the question of an "Asiatic mode of production," and probably on other aspects of the difficulties of fitting China's ancient history into Marxian boxes. (6,300)

1.2.4 Wang Chung-lo 王仲荦, Kuan-yü Chung-kuo nu-li she-hui ti wa-chieh yü feng-chien kuan-hsi ti hsing-ch'eng wen-t'i 関于中国奴隶社会的瓦解与封建関係的形成问题 (On the problem of the dissolution of the Chinese slave society and the formation of feudal relations; Wuhan: Hu-pei jen-min ch'u-pan-she, 1957), 135 pp.

This volume combines two articles originally published in Wen shih che in 1954 and 1956. The author's view is that "slave society" in China continued from the Shang through the Eastern Han dynasties (ca. 1760 B.C.-A.D. 190) and that the feudal period began in the Wei and Chin dynasties (A.D. 220-420). Within this scheme, he discusses the structure of the slave society, the decline of rural "communes," the strengthening of private slave-owning during the Han, the appearance of feudalistic "elements," the "crisis" of slave society, and the formation of feudal relations (a process in which he ascribes an important role to the "military agricultural colonies" [tun-t'ien 屯田] of Ts'ao Ts'ao). This is an almost exact copy of the usual Marxist picture of European economic and social development. (28,500)

1.2.5 Chinese History Seminar, People's University, comp., Chung-kuo nu-li-chih ching-chi hsing-t'ai ti p'ien-tuan t'an-t'ao 中国奴隶制經济形态的片断探討 (Preliminary explorations into the economic structure of the Chinese slave system; Peking: San-lien shu-tien, 1958), 13 + 329 pp.

This volume contains seven articles written by members of the Chinese History Seminar, brought together with a preface by Professor Shang Yüeh 尚鉞 as an investigation into the periodization of ancient Chinese history. While taking somewhat different approaches, the articles are all based on the premise that the Han dynasty was a slave society. The titles are: (1) "Did Chinese feudal society begin with the Western Chou?"; (2) "An exploration into the mode of production in the early Ch'in"; (3) "Commodity production and commerce in the two Han periods"; (4) "An exploration of the question of the nature of society in the two Han periods (being also a critique of Chien Po-tsan's 翦伯贊 'Concerning the question of public and private slaves in the two Han periods')"; (5) "A further discussion of the Han as a slave society"; (6) "On the question of the collapse of the Chinese slave society"; (7) "An exploration of the doctrine and ideology of Wang Ch'ung 王充 (A.D. 27-97)." (2,500)

1.2.6 Ts'en Chung-mien 岑仲勉, Hsi-chou she-hui chih-tu wen-t'i 西周社会制度問題 (Questions concerning the Western Chou society; Shanghai: Hsin-chih-shih ch'u-pan-she, 1956), 176 pp. (?)

1.2.7 Shu Shih-cheng 束世澂, *Chung-kuo ti feng-chien she-hui chi ch'i fen-ch'i* 中国的封建社会及其分期 (China's feudal society and its periodization; Shanghai: Hsin chih-shih ch'u-pan-she, 1957), 2 + 95 pp.

Two essays, first published in 1955 in the *Hua-tung shih-ta hsueh-pao* 华东师大学报 (Journal of the East China Normal University), are here reprinted with revisions. The first (pp. 1-42) argues that China's feudal society came into being during the Western Chou. Although many sources are cited, the reasoning is based on Marxism in its most simplified form. The second (pp. 43-94) is more interesting in that it is one of the relatively few published attempts to "periodize" the more than 2000 years of "feudalism" that allegedly followed the Chou. Although mechanically done and relying heavily on Soviet periodization discussions, it does recognize the fact that all was not simply unchanging "feudalism" from 770 B.C. to the nineteenth-century European impact. (12,000)

1.2.8 *Chung-kuo chin-tai shih fen-ch'i wen-t'i t'ao-lun chi* 中国近代史分期問題討論集 (Collected essays on the problem of the periodization of modern Chinese history), compiled by the editors of *Li-shih yen-chiu* 历史研究 (Historical studies; Peking: San-lien shu-tien, 1957), 2 + 242 pp.

Sixteen articles published originally between 1954 and 1957 are reprinted in this collection (with occasional revisions). Their theme is how to "periodize" China's "modern history," that is, the years 1840-1919. An astonishingly large number of ways to cut the pie are proposed (e.g., seven stages: 1840-50, 1851-64, 1864-95, 1895-1900, 1901-05, 1905-12, 1912-19; or four stages: 1840-64, 1864-95, 1895-1905, 1905-19; and a half dozen other variations). These are all scrupulously related to such Marxist-Maoist variables as changes in the mode of production, the class struggle, or principal "contradictions" within Chinese society. Pages 233-242 contain a brief article summarizing the discussions as of the end of 1956. Although the results seem sterile, the specialist may be interested in some of the finer points that are discussed in passing. (17,000)

1.3 NON-CHINESE MINORITY NATIONALITIES

The mainland historians hold that Chinese historiography prior to 1949 either totally ignored or gravely slandered the history of the non-Chinese nationalities within the borders of the Chinese empire, who are now referred to as the "brother" nations. But early post-1949 treatment of this matter, despite a more positive evaluation of minority opposition to the Ch'ing, for example, continued to be colored by Chinese nationalism. Thus, the conquest of the Hsiung-nu tribes by Han Wu-ti was interpreted as a "progressive" act because it contributed to the inevitable advancement of history through the Marxist stages, because it was motivated by the desire to protect the superior and relatively peaceful class relations of the Han nation, because it was the Hsiung-nu ruling class which was hardest hit, and because through the conquest the more advanced Chinese culture was propagated among the Hsiung-nu. In the same vein, the revolt of Yakub Beg against the Ch'ing was not "progressive" because it resulted in increased exploitation of the Muslim people, and was in fact a tool of British imperialist expansion into Chinese Turkestan. Therefore, the Ch'ing suppression of Yakub Beg is to be judged positively. It seems manifest that the increased attention given to the history of non-Chinese minorities is partly motivated by a desire to tie them closely to Han China and to reassert and underline Chinese sovereignty over her border areas.

1.3.1 Chien Po-tsan 翦伯贊 et al., eds. Li-tai ko-tsu chuan-chi hui-pien 歷代各族傳記会編 (A compendium of the histories of minority nationalities in the successive dynasties; Peking: Chung-hua shu-chü, 1958), 4 parts, ? vols. (Part I, 1 vol., and Part II, 2 vols. seen).

This large compilation consists of the "biographies" (chuan-chi) of non-Chinese minorities and "foreign countries" bordering on China extracted from the twenty-four dynastic histories (T'ung-wen ed.) and the Ch'ing shih kao (1927 ed.), with the addition of copious notes derived from commentaries, textual studies, and the like. Chien Po-tsan's preface claims that there is a great need for research and publication on the history of the racial minorities of China for the edification of both the Han Chinese and the minorities. He also explains that the inclusion of biographies of "foreign countries" along with those of the minority races is deemed necessary in view of the shifting Sino-barbarian frontier. What is listed as a tribal or racial group within the Chinese empire for one dynasty may appear as a foreign country in the next dynastic history, and fice versa; thus certain minorities would be passed over if the "biographies of foreign countries" were completely excluded. These materials have undergone considerable editorial processing, including punctuation, the addition of parenthetical Western dates (year), deletion of repetitious material, and the like. Part I covers Shih chi, Han shu and San-kuo chih; Part II, Chin shu and the Northern and Southern dynasties; Part III, T'ang, the Five Dynasties, Sung, Liao, Chin, and Yüan; Part IV, Ming and Ch'ing. (2,300)

1.3.2 Huang Yuan-ch'i 黃元起, Chung-kuo li-shih shang min-tsu chan-cheng ti p'ing-p'an wen-t'i 中国历史上民族战争的評判

问题 (The question of evaluating national wars in Chinese history; Shanghai: Shang-hai jen-min ch'u-pan-she, 1956), 48 pp.

This small volume claims to examine the endemic conflicts between the Han Chinese on the one hand and, on the other, the non-Chinese national minorities and tribal groups on the northern frontier, by applying Marxist theories to concrete facts. Thus tribal and racial wars are defined as a product of class society, a manifestation of class contradictions, and the wars themselves are either "righteous" or "unrighteous." But despite the claim of a more positive evaluation of minority opposition to Chinese rule than was allegedly the case before 1949, the CCP treatment of this conflict remains colored by Chinese nationalism. (8,000)

1.3.3 Pai Shou-i 白寿彝 et al., Hui-hui min-tsu ti li-shih ho hsien-chuang 回回民族的历史和現狀 (The history and present situation of the Muslim peoples [in China]; Peking: Min-tsu ch'u-pan-she, 1957), 63 pp.

This small volume is based in part on Hui-hui min-tsu wen-t'i 回回民族問題 (The question of the Muslim peoples), prepared by the Nationalities Research Association (Min-tsu wen-t'i yen-chiu hui 民族問題研究会) at Yenan during World War II. The historical survey is rather sketchy, with emphasis on the "patriotic" activities of the Chinese Muslims, first against the Ch'ing and later against the Japanese. The treatment of Muslim achievements under Communist rule (pp. 44-63) is along the same line. (15,000)

1.3.4 Ma Shao-ch'iao 馬少桥, Ch'ing-tai Miao-min ch'i-i 清代苗民起义 (Uprisings of the Miao people in the Ch'ing period;

Wuhan: Hu-pei jen-min ch'u-pan-she, 1956); 76 pp., 1 plate.

A semi-popular account of the rebellions of the Miao people in Kweichow and Hunan against the Ch'ing authorities in three periods: 1735-36, 1795-1806, and 1855-72. There is an introductory discussion of the legendary past of the Miao tribes, their geographic distribution, their society, and the oppressive policy of the Ch'ing government. The author holds that the chief cause of these rebellions was the replacement of hereditary chieftains with government-appointed officials, by which means the Ch'ing government confiscated land, levied taxes, and established military colonies on Miao lands. The author sees a correlation between the success of the Taiping movement and the Miao risings in the nineteenth century, and finds comparable reasons for their failures. Ch'ing official documents and local gazetters have been used as the main sources. (16,000)

1.3.5 Lin Kan 林幹, Ch'ing-tai Hui-min ch'i-i 清代回民起义 (Muslim uprisings in the Ch'ing period; Shanghai: Hsin chih-shih ch'u-pan-she, 1957); 73 pp.

This small book deals with the rebellions of (1) the Muslims of Kansu in 1648, (2) the Salar Muslims (of Turkish origin, also in Kansu) in 1781 and again in 1783, and (3) the series of risings in Shensi, Kansu, and Chinghai during the Taiping period, between 1862 and 1873. For (1) and (2), no sources are given; a number of direct quotations in (3) are taken from the four-volume Hui-min ch'i-i, ed. Pai Shou-i (1.3.6). The "proper understanding" of these uprisings, the author states, requires that they be seen both as national struggles and as revolts of the peasantry against the landlord-dominated feudal Ch'ing government. As a conclusion, the tradition of Muslim uprisings is linked with the liberation of the Chinese people by the CCP. (11,000)

1.3.6 Pai Shou-i 白寿彝, ed., Hui-min ch'i-i 回民起义 (The Muslim rebellions), Modern Chinese Historical Materials Series, No. 4, (Shanghai: Shen-chou kuo-kuang she, 1952), 4 vols., 2,001 pp.

The core of the first two volumes of this collection consists of documents, many of Muslim origin, previously printed in Professor Pai's Hsien t'ung Tien-pien chi-wen (2 vols., Chungking: Commercial Press, 1945). These are concerned with Muslim uprisings in Yunnan and Kweichow in the mid-nineteenth century. Of the sixty items in Vols. 1 and 2, sixteen are new and thirty-seven are rare. Vols. 3 and 4 contain materials on Muslim uprisings in northwest China, a much larger proportion of which are well-known to scholars than is the case with the contents of Vols. 1 and 2. For each item in this collection, the editor has supplied very useful introductory notes. There is no bibliography such as is found in other titles in this series. For a detailed review, see The Journal of Asian Studies, 17.1:80-86 (Nov. 1957). (7,200)

1.3.7 Ch'iao Pa-shan 喬巴山 [Choibalsang], Meng-ku jen-min ko-ming chien-shih 蒙古人民革命 (Concise history of the Mongolian people's revolution; Peking: Shih-chieh chih-shih she, 1956), 71 pp.

This is a Chinese translation, based on a Russian translation (published in 1952) of a book written by Mongolia's late premier, Marshal Choibalsang, who was a political leader and military commander during Mongolia's "struggle for independence" between 1920 and 1921. This short, undocumented history touches on conditions in Mongolia during its nominal autonomy (1911-19), the removal of its autonomous status by China in 1920, the rise of the revolutionary movement and its contact with the Soviet Union, the defeat of the White Russian forces of Baron Ungern-Sternberg, and the formation of the Mongolian People's Revolutionary Government under Soviet auspices in 1921. A number of explanatory notes have been added by both the Russian and Chinese translators for names that are not identified in the text. (10,000)

1.4 SURVEYS OF MODERN AND CONTEMPORARY HISTORY

Chinese Communist historiography takes the Opium War as the starting place for modern history. There is general agreement that in the Opium War and its aftermath Chinese "feudal" society began to be transformed into "semi-feudal, semi-colonial" society, which in turn lasted until the establishment of the People's Republic in October 1949. The last century is divided as follows: "Modern history" (chin-tai shih 近代史) extends from 1840 to 1919 and is characterized by the gradual development of the forces of the "Old Democratic Revolution" (or bourgeois revolution) which reached its culmination in the republican movement led by Sun Yat-sen. The years 1919-49 are the period of the "New Democratic Revolution" under the leadership of the Chinese Communist Party, and are referred to as "contemporary history" (hsien-tai shih 現代史). Current history or the "Epoch of the People's Republic of China" begins with 1949. The content which is fitted into this historical outline is summarized in Mao's statement: "The history of imperialist aggression upon China, of imperialist opposition to China's independence and to her development of capitalism, constitutes precisely the history of modern China. Revolutions in China failed one after another because imperialism strangled them."

For self-professed "scientific" historians, however, the outlines of this foreign imperialist evil which dominates modern history are remarkably shadowy and fluid. It is notable to what degree historical narration in practice tends to lump all foreign contact with China in the modern period under the heading "imperialist aggression" and to ignore any finer distinctions. A prime example of this tendency is Mao himself, in whose doctrinal writings "foreign capitalism" and "foreign imperialism" are often loosely interchangeable. The contrast between the specific formulations of Lenin and the vagueness of Chinese Communist historians is so marked that we can hardly believe that the Chinese are not aware of it. It would be misleading to interpret the protean and omnipresent application of the touchstone "imperialism" in current mainland writing as a serious attempt to employ the tools of Marxism in attempting to comprehend the history of the past century. Rather, the study and writing of modern Chinese history in the People's Republic of China at the present time tends to be largely an ideological exercise. As such, it is designed to harness and channel the real political and economic frustrations encountered in China's nineteenth- and twentieth-century experience, in the interests of the new historical integration under the auspices of the Chinese Communist Party. With the outline and content of modern history thus so fully prescribed by the political leaders of the regime, the historians have been able to do no more than turn with a vengeance to the task of elaborating the stages and sub-stages of the past century of China's confrontation with the West.

The first five items in this section are relatively scholarly surveys of modern history. Item 1.4.1 is remarkable in being an old-style compilation with hardly a trace of the Marxist-Maoist doctrine. Items 1.4.8-19 are the principal examples of the more popular and propagandistic variety of survey, arranged here in order of their publication. These are followed by several compilations of sources and reference materials, which in general are less valuable than the more detailed documentary collections treating specific topics in modern history which are discussed in Section 2 of this volume. Items 1.4.24-31 treat in turn of Sino-Soviet relations and Sino-American relations. It is, of course, American "imperialism" which is the chief target of the mainland writers on history. The result is a picture of Chinese-American relations which will not be recognizable to any American student of this topic. On this subject see also 2.9.15-18, and 1.4.4, the first volume of a projected large-scale study of imperialist aggression against China.

1.4.1 Teng Chih-ch'eng 鄧之誠, Chung-hua erh-ch'ien-nien shih 中華二千年史 (Two thousand years of China), Part 5, 3 vols. (Peking: Chung-hua shu-chü, Vol. 1, 1956; Vols. 2-3, 1958), 1742 pp.

The publication of these volumes on the Ming and Ch'ing completes Professor Teng's monumental history of China begun some twenty years ago. (Parts 1-4, which come up to the Ming dynasty, were published by the Commercial Press in 1934, and were reprinted, with minor revisions, by Chung-hua shu-chü in 1954.) The style and form of Part 5 are similar to those of the previous parts, but the treatment is much more detailed. The narrative, which is in the literary style, is frequently interspersed with passages taken from the dynastic histories and other sources, and is accompanied by a number of tables (e.g., table of countries which had maritime relations with the early Ming dynasty; table of cases involved in the Ch'ing literary inquisition). The topical and chronological treatment of the two dynasties (1644-1911) is followed by a section devoted to Ming and Ch'ing institutions (e.g., the system of landholding, taxation, corvée, currency, the examination system, the army, the legal system, technology, commerce, industry, scholarship, literature). The twenty-year-old draft does not appear to have been altered in any way to take account of the new historiographical doctrines. (6,800)

1.4.2 Fan Wen-lan 范文瀾, Chung-kuo chin-tai shih 中国近代史 (Chinese modern history; Peking: Hsin-hua shu-tien, 1949), Part A, Vol. 1, 9 + 543 pp.; another edition published by Jen-min ch'u-pan-she, 1952, 8 + 454 pp.

This volume (first published in 1947 under the same title but with the author's pseudonym, Wu P'o 武波), describes the "formation of China's semi-feudal, semi-colonial society and the people's old-style resistance movement," from 1840 to 1905, as part of a survey of China's modern history from the Opium War of 1840 to the May Fourth Movement of 1919. It was one of the first Communist interpretations of the modern period; and while in certain respects it is now outdated, it remains influential. Later editions contain additional sections and revisions. The projected second volume has never appeared. (74,000)

1.4.3 Tai I 戴逸, <u>Chung-kuo chin-tai shih kao</u> 中国近代史稿 (Draft history of modern China; Peking: Jen-min ch'u-pan-she, 1958), Vol. 1, 4 + 523 pp.

This is the first volume of what is intended to be a comprehensive university-level text for modern history. Its five chapters treat the following subjects: (1) the world situation before 1840 and China; (2) the first Opium War and the beginning of China's "semi-colonization"; (3) the first stage of the Taiping Rebellion (to 1856); (4) the second Opium War, the Taiping Rebellion, and other uprisings; (5) political and economic factors in the Taiping Rebellion, the reasons for its failure, and its historical significance. The text is based on the author's lectures at the Chinese People's University, Peking. Included are seven maps and numerous footnotes referring to major sources for the period. (10,000)

1.4.4 Ting Ming-nan 丁名楠, et al., <u>Ti-kuo-chu-i ch'in Hua shih</u> 帝国主义侵华史 (A history of imperialist aggression against China; Peking: K'o-hsueh ch'u-pan-she, 1958), Vol. 1, vi + 333 pp., plates.

The first volume of a detailed history of China's foreign relations in the nineteenth and twentieth centuries, the major theme of which is China's descent into a semi-colonial status as a consequence of foreign capitalist-imperialist aggression. Vol. 1 covers the period 1840-95 in an orthodox chronological fashion, is carefully documented, draws on a wide range of the best Chinese and Western-language sources, and does not deviate from what has become the orthodox CCP treatment of this period. (5,975)

1.4.5 Hu Sheng 胡繩, Ti-kuo-chu-i yü Chung-kuo cheng-chih 帝国主义与中国政治 (Imperialism and Chinese politics), 2nd ed. (Peking: Jen-min ch'u-pan-she, 1953), 5 + 222 pp.

This book was first published in 1948, and has been slightly revised for the present edition. It deals with the period 1840 to 1924, during which Chinese politics are seen as largely a reaction to the "imperialist" activities of the Western powers. The author's theme, developed with some skill, is the transformation of China into a "semi-colony." Throughout he emphasizes the relations between the "imperialist aggressors" and their political tools in China, and plays up the different attitudes of the "reactionary rulers" and the "people of China" towards imperialism. Widely distributed in Chinese and foreign language versions (e.g., Imperialism and Chinese Politics, Peking: Foreign Languages Press, 1955), Hu Sheng's text is probably the best of the recent works of its kind and political outlook. (53,000)

1.4.6 Li Shu 黎澍, Chin-tai shih lun ts'ung 近代史論叢 (Collected essays on modern history; Peking: Hsueh-hsi tsa-chi she, 1956), 136 pp.

The seven essays (previously published in periodicals and now revised) are entitled: (1) "The Russian Revolution of 1905 and China"; (2) "The spread of socialism in China"; (3) "The May Fourth Movement"; (4) "Sun Yat-sen's struggles after the failure of the 1911 Revolution"; (5) "On 'reformism' in modern Chinese history"; (6) "The true reactionary character of the so-called democratic politics of the Hu Shih faction"; and (7) "An inquiry into the question of incipient Chinese capitalism." According to the author's foreword, he is concerned with these three major themes: the nature and degree of Russian revolutionary influence on contemporary Chinese intellectual life, the reactionary function of the bourgeois reformists represented by Hu Shih, and an evaluation of incipient capitalism in Chinese feudal society. With regard to the last, Li Shu opposes Shang Yüeh (see sec. 4.3) and others who have ascribed capitalistic elements to the Ming dynasty. He sees the growth of capitalism in China as a much slower and later process. (28,100)

1.4.7 P'eng Ming 彭明, Chung-kuo chin-tai li-shih ku-shih 中国近代历史故事 (Historical tales of modern China; Shanghai: Commercial Press, 1950, 11th printing, 1956), 84 pp.

1.4.8 Hu Hua 胡华, Chung-kuo hsin min-chu-chu-i ko-ming shih, ch'u-kao 中国新民主主义革命史初稿 (A history of China's new democratic revolution, first draft; Peking [and elsewhere]: Hsin-hua shu-tien, 1950), 8 + 243 pp.

A popular textbook on "Chinese revolutionary history," which here means the years 1919-45. This is a field which has been taught increasingly in Communist Chinese schools. The narrative is divided into the usual main periods: the May Fourth Movement, the two civil wars, and the war with Japan. The treatment of the

Nationalist-Communist struggles between 1925 and 1937 is relatively detailed, but the text is not documented. A revised version was published by the Jen-min ch'u-pan-she in 1952. (342,000)

1.4.9 Hua Kang 華崗 , Chung-kuo min-tsu chieh-fang yun-tung shih 中国民族解放運動史 (A history of China's national liberation movements), Vol. 1, enlarged ed. (Peking: San-lien shu-tien, 1951), 11 + 606 pp.

This is a revised and expanded version of the first of two volumes originally published in 1940. The first edition, which was only half as long as the present volume, had been used as a textbook in Communist areas and had gone through many printings. (A revised edition of the second volume, though promised, has apparently not appeared.) The introductory chapter discusses "national liberation," i.e., revolutions, in a Marxist-Leninist framework, and also surveys the "formation of the Chinese nation" and premodern "racial conflicts and national movements." The text treats Chinese modern history by dividing it into the now standard periods: (1) Opium War, (2) Taiping Rebellion, (3) Second Opium War, (4) Sino-French War and Sino-Japanese War, (5) Reform of 1898 and Boxer Rebellion, (6) 1911 Revolution, and (7) May Fourth Movement. Within these periods the narrative is relatively detailed, particularly with regard to events and ideas that the author finds useful in advancing his Marxist-Leninist-Maoist thesis. (40,000)

1.4.10 Yeh Huo-sheng 葉蠖生 , Hsien-tai Chung-kuo ko-ming shih hua 現代中国革命史話 (An informal history of the contemporary Chinese revolution; Peking: Chung-kuo ch'ing-nien ch'u-pan-she, 1954, first pub. 1951), 8 + 110 pp. (275,000)

1.4.11 Shui Chao-hsiung 水兆熊, Chung-kuo chin-tai shih hsueh-hsi kang-yao 中国近代史学習綱要 (An outline for the study of modern Chinese history; Szechwan: T'ai-lien ch'u-pan-she, 1952), 8 + 116 pp.

A textbook; from the First Opium War down to the warlord period. (3,000)

1.4.12 Chang Shou-ch'ang 張守常, Chung-kuo chin-tai shih kang-yao 中国近代史綱要 (An outline of modern Chinese history; Tientsin: Li-shih chiao-hsueh yueh-k'an she, 1952), 3 + 110 pp.

A textbook for senior-middle school students. (3,500)

1.4.13 Liu P'ei-hua 刘培華, Chung-kuo chin-tai chien-shih 中国近代簡史 (Concise history of modern China; Peking: I-ch'ang shu-chü, 1953), 10 + 253 + 18 pp., maps.

This book is a product of the author's classroom lectures at a middle school in Peking. It consists of nine units (the Opium War, the Anglo-French expedition, the Taiping movement, the yang-wu or "Westernization" movement, the Sino-French War, the Sino-Japanese War of 1894, the Reform Movement of 1898, the Boxer uprising, the 1911 Revolution, and the Peiyang warlord period), which indicate the major topics currently treated at the secondary school level. The text is accompanied by a number of fairly useful maps. (5,000)

1.4.14 Lu T'ien 陸天, Chung-kuo chin-tai shih hsueh-hsi wen-ta 中国近代史学習問答 (Questions and answers for the study of Chinese modern history; Shanghai: Shih-hsi ch'u-pan-she, 1953), 108 pp.

This is an example of a new genre, history for the newly-literate masses. The simple questions and answers on Chinese modern history are designed to give basic historical facts while stressing such standard Chinese Communist themes as capitalist aggression, Ch'ing decadence, and the people's patriotism. (6,000)

1.4.15 Meng Chün 孟均, Chung-kuo hsin mín-chu-chu-i ko-ming shih hsueh-hsi wen-ta 中国新民主主义革命史学習問答 (Questions and answers on China's new democratic revolutionary history; Shanghai: Shih-hsi ch'u-pan-she, 1953), 150 pp. (6,000)

1.4.16 Jung Meng-yuan 榮孟源, Chung-kuo chin-pai-nien ko-ming shih lueh 中国近百年革命史略 (A brief history of China's revolutions in the past century; Peking: San-lien shu-tien, 1954), 6 + 233 pp.

This brief textbook offers one more example of the standard Chinese Communist interpretation of Chinese history from 1830 to 1949. It is organized as follows. Chapter 1: Because the "contradiction" between imperialism and the Chinese nation is the "principal contradiction" out of which China's great revolution arose, the "historical facts" of imperialist aggression toward China introduce the book. Chapter 2: The Taiping revolution was an anti-imperialist and anti-feudal struggle which sprang up among the peasants. Chapter 3: The 1898 Reform was a movement of intellectuals of landlord class origin and capitalist ideology. Chapter 4: The Boxer incident--same class origin as the Taipings. Chapter 5: The 1911 Revolution--not the "real revolutionary storm," but it gave an impetus to demoncratic revolutionary movements. Chapters 6-10: The "real revolution" led by the proletariat: the May Fourth Movement, the two civil wars (i.e., 1921-27 and 1928-37), the war with Japan, and the third civil war (i.e., 1945-49), culminating in the people's victory. The conclusion reiterates what "the history of China's last hundred years proves": revolutionary success = proletarian leadership + Marxism-Leninism + Maoism. (100,000)

1.4.17 Liang Han-ping 梁寒冰, Chung-kuo hsien-tai ko-ming shih chiao-hsueh ts'an-k'ao t'i-kang 中国現代革命史教学参考提綱 (An outline for instruction in the history of revolutions in contemporary China; Tientsin: T'ien-chin t'ung-su ch'u-pan-she, 1955), 4 + 206 pp.

The basic content and organization of this widely circulated textbook are dervied from Hu Ch'iao-mu's Thirty years of the Communist Party of China (3.3.3), with material added from Mao's collected writings and other CCP literature. The manuscript was developed from the author's lectures at the "Marxist-Leninist Night School" for higher and middle school teachers, organized in Tientsin in 1954. It is divided into five parts which give a good indication of its emphasis: the formation of the CCP, the first and second revolutionary civil wars (1921-27 and 1927-37), the war of resistance against Japan, and the third revolutionary civil war (1945-49). (200,000)

1.4.18 Wang Po-yen 汪伯岩, Chung-kuo chin-tai shih chiang-hua 中国近代史講話 (A narrative on Chinese modern history; Tsinan: Shan-tung jen-min ch'u-pan-she, 1957), 4 + 258 pp.

This is a conventional, semi-popular, doctrinaire account of the period from 1840 to 1919, which had a very wide sale in its original form of six separate pamphlets published between 1954 and 1956. Like many other mainland books on Chinese modern history, it is organized around the Opium War, the Taiping Rebellion, the

Sino-Japanese War, the Reform Movement of 1898, the 1911 Revolution, and finally, the warlord period. (54,100)

1.4.19 Ho Kan-chih 何幹之, chief ed., Chung-kuo hsien-tai ko-ming shih 中国現代革命史 (A history of the contemporary Chinese revolution; Hong Kong: San-lien shu-tien, 1958), 5 + 398 pp.

In this undeviatingly party-line textbook, the history of the modern Chinese revolution is defined as "the history of the struggles of the Chinese people under the leadership of the Chinese Communist Party for the fulfillment of the new democratic revolution and the construction of a socialist society." The fifteen chapters begin with the May Fourth Movement and continue down to 1956. Their chief virtue lies in the fact that they probably constitute the most detailed single Communist account of this period, and as such can serve as a guide to the current official line on these critical years of modern Chinese history. No indication is given of the sources used, except that the frequent quotations from Mao are carefully identified. Doctrinally the most important thing about this text is the fact that Stalin's role in the Chinese revolution is completely ignored, a trend already apparent in Hu Ch'iao-mu's briefer treatment of this period (3.3.3) and in sharp contrast with, for example, Ch'en Po-ta's Stalin and the Chinese Revolution (Peking: Foreign Languages Press, 1953). Stalin is mentioned only two or three times, while all the praise goes to Mao. A preliminary version was published in 1955, edited by Ho Kan-chih from "Lectures on Chinese Modern Revolutionary History" compiled by faculty members of several Peking colleges. A revised and enlarged edition in two volumes was issued in 1957 in Peking by the Higher Education Publishing House (Kao-teng chiao-yü ch'u-pan-she) in a printing of 185,000 copies. This Hong Kong edition is identical with this last. An English translation was published in Peking in 1959 by the Foreign Languages Press under the title *A History of the Modern Chinese Revolution*. (185,000)

1.4.20 Yang Sung 楊松, Teng Li-ch'ün 鄧力群, eds., *Chung-kuo chin-tai shih ts'an-k'ao tzu-liao* 中国近代史参考資料 (Reference materials for Chinese modern history; Peiping: Hsin Chung-kuo shu-chü, 1949), 428 pp.

This edition is presumably one of the reprints of a book first published in Yenan in 1940 (see Jung Meng-yuan's remarks in the preface to *Chung-kuo chin-tai-shih tzu-liao hsuan chi*, item 1.4.22). It consists of selections from the writings of such historical personages as Lin Tse-hsü, Hung Hsiu-ch'üan, and K'ang Yu-wei, as well as Marx, Engels, and Lenin. Under seven headings, the selections begin with the first Opium War and end with the Boxer incident. The entire collection is rather thin and has been superseded by 1.4.22. A calendar of corresponding Chinese and Western dates for the Ch'ing dynasty is appended. (10,000)

1.4.21 Hu Hua 胡華, chief ed., Tai I 戴逸, and Yen Ch'i 彥奇, eds., *Chung-kuo hsin min-chu-chu-i ko-ming shih ts'an-k'ao tzu-liao* 中国新民主主义革命史参考資料 (Reference materials on the history of China's new democratic revolution; Peking: Commercial Press, 1951), 14 + 494 pp.

Excerpts and documents, from the period of the May Fourth Movement (1916 *et seq.*) down to 1945, including a large number of CCP documents. The Chinese text of one of these ("Manifesto of the Second National Congress of the CCP," pp. 69-84) was not available to C. Brandt, B. Schwartz, and J. K. Fairbank for their *A Documentary History of Chinese Communism* (1952). In selection and arrangement the present volume is designed to give the Communist story and interpretation. (91,000)

1.4.22 Chung-kuo chin-tai shih tzu-liao hsuan-chi 中国近代史資料選輯 (Selected source materials on Chinese modern history), originally ed. by Yang Sung 楊松 and Teng Li-ch'ün 鄧力群 ; re-edited by Jung Meng-yuan 榮孟源 (Peking: San-lien shu-tien, 1954), 9 + 830 pp.

Although this volume is stated to be based on Yang and Teng's Chung-kuo chin-tai shih ts'an-k'ao tzu-liao (1.4.20) and those two men are therefore billed as the original editors, a comparison of the two publications shows that the present volume as "re-edited" by Jung is quite different. It omits twenty-five out of the forty-nine items in the earlier work and adds sixty-seven others, making a total of ninety-one selections. Whereas the earlier volume stops at about 1901, the present edition goes up to the end of World War I. The inclusion of several articles from non-Chinese sources plus a wider selection from the corpus of Chinese historical documents and records make the current volume more comprehensive and useful. Note also that Jung Meng-yuan, who was severely criticized as a "rightist" in 1957, has deleted all the selections from Marx, Engels, and Lenin that appeared in the earlier edition. (55,000)

1.4.23 Peking Historical Museum, ed., Chung-kuo chin-tai shih ts'an-k'ao t'u-p'ien chi 中国近代史参考圖片集 (A collection of pictorial references for Chinese modern history; Shanghai: Shang-hai chiao-yü ch'u-pan-she, 1958), 3 vols., 152 + 157 + 165 pp.

These volumes are designed for middle school history courses. In accordance with the current mainland definition, Chinese "modern" history covers the period 1840-1919. Thus Vol. 1 includes the Opium War and the Taiping movement; Vol. 2, conditions between the 1870's and 1890's, the Sino-Japanese War, the 1898 Reform and the

Boxer movement; Vol. 3, the 1911 Revolution (actually the years 1901-18). A few pictures, illustrating Chinese society before the Opium War are reproduced from Keng chih t'u 耕織圖 (first published in the Sung dynasty) and T'ien-kung k'ai-wu 天工開物 (Ming dynasty). The bulk of the illustrations are from the Peking Historical Museum collection or private sources. A small number of reproductions are derived from Japanese- or English-language books. The plates are all in black and white, but the numerous excellent historical maps are in color. Depicting a wide range of objects, people, places, notable events, representative documents, and the like, this pictorial record is of considerable interest. It is regrettable that so many of the half-tone plates are of such a poor quality. (2,500)

1.4.24 Hsin-hua Current-Event Periodicals Press, ed., Chung-Su kuan-hsi shih-liao 中苏関係史料 (Historical Materials on Sino-Soviet relations; Shanghai: Hsin-hua shu-tien, 1950), 2 + 77 pp. (?)

1.4.25 P'eng Ming 彭明, Chung-Su jen-min yu-i chien-shih 中苏人民友誼簡史 (Concise history of the friendship of the Chinese and Soviet peoples; Peking: Chung-kuo ch'ing-nien ch'u-pan-she, 1955), 4 + 151 pp.

A semi-popular account of the relations between the Soviet Union and the "progressive" forces in China from the October Revolution to 1953. The principal sources cited in footnotes are contemporary Chinese newspapers. A few samples of the section headings will indicate the general nature of the contents: "The Soviet people's support of the Chinese people during the May 30th Movement," "The Soviet people's support of the Chinese people's war of liberation," etc. (60,000)

1.4.26 P'eng Ming 彭明 , Chung-Su yu-i shih 中苏友誼史 (A history of Sino-Soviet friendship; Peking: Jen-min ch'u-pan-she, 1957), 296 pp., 7 plates.

This is a revised and enlarged edition of the previous item, which deals with Sino-Soviet relations from 1917 to the present. The author keeps his account focused on the friendly aspects of the relationship, and stresses China's indebtedness to the Russians for ideological, political, economic, and other help. Such touchy matters as the Chinese Eastern Railway question, or Mongolia, are skirted; and exclusive attention is given to the official public record. (14,000)

1.4.27 Liu Ta-nien 刘大年 , Mei-kuo ch'in Hua shih 美国侵华史 (A history of American aggression against China; Peking: Jen-min ch'u-pan-she, 1951), 7 + 252 pp.

Mr. Liu, associate editor of Li-shih yen-chiu, divides the history of Sino-American relations into four stages: (1) the U.S. follows other nations, or uses them as a medium, in carrying out aggression against China, 1840-1905; (2) the U.S. on the road to independent aggression, 1905-17; (3) the U.S. endeavors to gain control of China, 1917-45; (4) the U.S. attempts sole occupation of China, but fails, 1945-50. Within these periods the discussion of specific events is footnoted with references to Chinese and American official documents, monographs, and articles, as well as to a number of Japanese works. For the post-1945 period, the principal source consists of CCP newspapers. Throughout, the role and motives of the U.S. are handled in the expected doctrinaire manner, while the tone of the writing is decidedly propagandistic. (20,000)

1.4.28 T'ao Chü-yin 陶菊隱, Mei-kuo ch'in Hua shih-liao 美国侵华史料 (Historical material on America's aggression against China; Shanghai: Chung-hua shu-chü, 1951), 196 pp.

The purpose of this book, according to the introductory remarks printed on the inside of the cover, is to "strengthen our countrymen's psychology of regarding America as our enemy and despising it." It is further stated that the "language is common and fluent and rich in interest, and can be read as fiction." The so-called historical material consists of a highly selective narrative of Sino-American relations from the Opium War to the Korean War. American treachery is a foregone conclusion, and is frequently linked up with the actions of such Chinese figures as Li Hung-chang, Yuan Shih-k'ai, and Chiang Kai-shek. (12,000)

1.4.29 Peking Branch of the Chinese People's Committee for the Preservation of World Peace and Against American Aggression, ed., Mei-kuo ch'in Hua shih-liao 美国侵华史料 (Historical materials on American aggression against China; Peking: Jen-min ch'u-pan-she, 1951), 3 + 313 pp. (20,000)

1.4.30 "Historical Studies Weekly" Society, ed., Mei ti-kuo-chu-i ching-chi ch'in Hua shih lun ts'ung

(Collected essays on American imperialism's economic aggression against China; Peking: San-lien shu-tien, 1953), 2 + 129 pp. (8,000)

1.4.31 Ch'ing Ju-chi 卿汝楫, Mei-kuo ch'in-lueh T'ai-wan shih 美国侵略台湾史 (A history of American aggression against Taiwan; Peking: Chung-kuo ch'ing-nien ch'u-pan-she, 1955), viii + 106 pp. (140,000)

2. THE MING AND CH'ING DYNASTIES

The section which follows is the largest in this bibliography and contains the solid core of published volumes on modern history up to 1911 produced in the past decade in the People's Republic of China. The works discussed below include collections of documents, specialized monographs, and semi-popular textbooks. As a group they come most easily under the rubric "political history." Works concerned primarily with economic history and the history of thought are discussed in Sections 4 and 5 respectively.

It is of some interest, we believe, that in the topics covered, although not in their treatment, the mainland historians have departed very little from the chief concerns of Chinese historiography in the 1930's. This observation should perhaps be qualified by pointing out the increased emphasis on such factors as peasant rebellions and the role of imperialism, but the general impression remains that Chinese Communist historiography in the past decade has not been outstandingly creative in searching out new problems or suggesting broad new concepts for the study of modern Chinese history. The bulk of the works in this part deal with such time-worn topics as the Opium War, the Taiping Rebellion, the Reform Movement of 1898, and the Boxer Rebellion. There has been little effort to fill in the interstices between these highlights, or to connect them in anything other than a crude Marxist framework. The most valuable part of the historical output of the past decade is clearly the assiduous collection and publication of source materials, many of them not otherwise available outside of mainland China. If these documentary volumes are tendentious in their arrangement, their value to the Western scholar of modern Chinese history is perhaps reduced to that extent, but by no means canceled.

2.1 MING HISTORY

Materials relating to the history of the Ming Dynasty (1368-1644) appear also in works discussed in sec. 1.1 (on peasant uprisings), in secs. 4.1-4 (on economic history), and in secs. 5.1 and 5.5-6 (on intellectual and cultural history).

The early Ming, especially the reign of Chu Yuan-chang, is in general treated with respect by the mainland historians, perhaps a reflection of their nationalistic reaction to the preceding period of Mongol rule. By the late Ming, it is held, Chinese feudal society had begun to decay internally and elements of incipient capitalism had begun to appear within the feudal nexus. China would have developed slowly into a capitalist society, had not the Manchu invasion and devastation of the land set the process back by two centuries or more and prevented this happy fruition.

Item 2.1.1 is a brief survey of Ming and early Ch'ing history which summarizes the current interpretation. Compare the traditional account of Teng Chih-ch'eng (1.4.1, which is included in sec. 1.4 because it carries the history of the Ch'ing Dynasty down to 1911). In a like manner, 2.1.2 and 2.1.3 are contrasting Marxist and traditional treatments of a wide variety of topics in Ming and early Ch'ing history. The next six entries (2.1.4-9) are reprints of basic sources on Ming history, 2.1.5 being of particular importance because it had existed previously only in manuscript form. A large number of reprints of Ming and Ch'ing works of the *pi-chi* genre have been announced, but we have seen only those included here and in sec. 2.3. Items 2.1.10-11 deal largely with the baleful role of eunuchs, and with their opponents, in the Ming Dynasty. They are followed by studies of three important Ming figures, including Wu Han's biography of Chu Yuan-chang. Items 2.1.20-24 contain materials on peasant uprisings at the end of the Ming Dynasty and an account of the last resistance of the Ming against the Manchu invaders.

2.1.1 Li Hsün 李洵, <u>Ming Ch'ing shih</u> 明清史 (A history of the Ming and Ch'ing dynasties; Peking: Jen-min ch'u-pan-she, 1956), 261 pp., 16 plates.

The main periodization scheme worked out by the author is as follows: (1) 1360-1430: recovery of Chinese society and economy from Yuan "barbarian" control, and continued development; (2) 1430-1560: crisis of Chinese feudal society; (3) 1560-1640: appearance of incipient capitalism in Chinese feudal society; (4) 1640-1780: consolidation and expansion of Manchu rule, and retardation of the growth of the capitalist economy; (5) 1780-1840: gradual disintegration of feudal economy, growth of capitalist economic elements, rise of nationalist opposition to the Manchus, and the first incursion of foreign capitalism. Within this general framework, the chapters and subsections are buttressed by frequent references to the standard Ming and Ch'ing histories and documentary collections. Originally put together as the author's lecture notes to his university classes, this book lists several themes for discussion at the end of each chapter, e.g., "the real nature of the 'single whip system' and the social background of its appearance" (p. 119). There is also a brief discussion of the sources and historiography of the period (pp. 11-18). (20,000)

2.1.2 Li Kuang-pi 李光璧, ed., <u>Ming Ch'ing shih lun ts'ung</u> 明清史論叢 (Collected essays on Ming and Ch'ing history; Wuhan: Hu-pei jen-min ch'u-pan-she, 1957), 294 pp.

Eighteen articles, for the most part reprinted from the magazine for teachers of history, <u>Li-shih chiao-hsueh</u> (6.4.16), are collected in this volume. Some are by authors such as Li Wen-chih 李文治 and Liang Fang-chung 梁方仲, who are acknowledged specialists on the Ming or Ch'ing periods. The topics include: (1) the land and land tax systems, (2) agriculture and handicraft industry, (3) peasant uprisings, (4) resistance of Ming loyalists to the Ch'ing, (5) developments in science and technology. Liang Fang-chung's article on the Ming "land-tax collector"system (liang-chang 粮長) has been further expanded and published as a book (4.2.2). An essay on Paul Hsü (Hsü Kuang-ch'i 徐光啓) and another on the late Ming encyclopedia of productive technology, T'ien-kung k'ai-wu 天工開物 , are published here for the first time. (13,500)

2.1.3 Meng Sen 孟森, Ming Ch'ing shih lun chu chi-k'an 明清史論著集刊 (Collected essays on Ming and Ch'ing history; Peking: Chung-hua shu-chü, 1959), 2 vols., 6 + 634 pp.

Forty-four articles by Professor Meng Sen (1867-1937) are collected in these two volumes. Only a small number are published here for the first time; the majority have appeared before in periodicals or in his earlier collected works. Besides studies of historical events and institutions of the Ming and Ch'ing dynasties, there are also textual studies and commentaries on historical materials, mostly of the Ch'ing period. All the articles are in the literary style, with punctuation added. Although the editor comments on the author's "feudal ideology," Meng Sen, an "old-model historian," is apparently still enjoying some esteem on the mainland (as well as on Taiwan, where his Ming-tai shih 明代史 [History of the Ming dynasty] was reprinted in 1957). About fifty other articles by Professor Meng not included in this collection are listed in an appendix. (3,100)

2.1.4 Lung Wen-pin 龍文彬, comp., Ming hui-yao 明會要 (Collected statutes of the Ming dynasty; Peking: Chung-hua shu-chü, 1956), 2 vols., 24 + 1563 pp.

This collection of Ming documents in eighty chüan by a Ch'ing editor is reprinted in a handsome, punctuated edition, without any introductory remarks by the publisher. (4,000)

2.1.5 T'an Ch'ien 談遷, Kuo chüeh 國榷 (Ming annals) collated by Chang Tsung-hsiang 張宗祥 (Peking: Ku-chi ch'u-pan-she, 1958), 6 vols., 17 + 6217 pp.

Written about 1630, these annals of the Ming dynasty existed only in manuscript form until the present printing. Two manuscript versions were used by the collator-editor, whose preface contains information about the author and about several incomplete manuscripts of this work. As historical material on the Ming dynasty, Kuo-chüeh is considered particularly valuable for its treatment of Ming relations with the pre-Ch'ing Manchu state of Chien-chou 建州 during the Wan-li period (1573-1619) and later. (1,000)

2.1.6 Hsia Hsieh 夏燮, Ming t'ung-chien 明通鑑 (A history of the Ming dynasty; Peking: Chung-hua shu-chü, 1959), 4 vols., 34 + 3765 pp.

Based on an edition of 1897 (published in Hupei), this reprint is punctuated and has a modern format. The publisher's note contains some introductory and critical remarks about the work (written shortly after 1860), pointing out that the patriotic spirit that permeates it is worthy of attention. The total of 100 chüan are in four volumes of large print. (1,100)

2.1.7 T'ao Tsung-i 陶宗儀, Nan-ts'un cho-keng lu 南村輟耕錄 (Passages written between farming labors), in Yuan Ming shih-liao pi-chi ts'ung-k'an 元明史料筆記叢刊 (Yuan and Ming historical materials and notes series; Peking: Chung-hua shu-chü, 1959), 13 + 385 pp.

A reprint of a much-reprinted book, based on a 1923 edition. As the title implies, the author was a "gentleman farmer" born at the end of the Yuan dynasty, who allegedly did his writing on tree leaves, which he kept in big pots, over a period of ten years. The contents, originally divided into thirty chüan, consist of numerous entries of varying lengths on a miscellaneous assortment of topics (e.g., genealogy of the Ming royal family, descriptions of mythical creatures and plants, sources of the Yellow River, names of famous physicians over several dynasties, painting techniques, and musical instruments). As usual, there is no index, even in this modern edition. (1,600)

2.1.8 Shen Te-fu 沈德符, Wan-li yeh huo pien 萬曆野獲編 (A collection of "wild harvests" of the Wan-li period), in Yuan Ming shih-liao pi-chi ts'ung-k'an (Peking: Chung-hua shu-chü, 1959), 3 vols., 31 + 938 pp.

The author (1578-1642) was a chü-jen of the Wan-li period, who lived from childhood until middle age with his father and grandfather in Peking, where they were both officials. This book is a collection of tales about the Ming court which the author heard from his father and grandfather, plus other topics he picked up elsewhere. The present reprint, based on an 1827 edition, consisting of thirty-four chüan and forty-eight main categories, including the court, the cabinet, women in the palace, imperial relatives, eunuchs, examinations, the Six Boards, and the imperial guards. (1,300)

2.1.9 Ho Liang-chün 何良俊, Ssu-yu-chai ts'ung-shuo 四友齋叢說 (Collected notes from the Ssu-yu Studio [containing historical material on the Ming]), in Yuan Ming shih-liao pi-chi ts'ung-kan (Peking:

Chung-hua shu-chü, 1959, first publ. 1569), 12 + 350 pp. (2,100)

2.1.10 Ting I 丁易, Ming-tai t'e-wu cheng-chih 明代特務政治 (The Ming dynasty's government by secret police; Peking: Chung-wai ch'u-pan-she, 1950), 10 + 617 pp.

Although the author's intensely negative value judgments are very forcefully stated, this volume is still useful as a well-documented, detailed, descriptive treatment of the peculiarities of the Ming government--its power hierarchy, land-holding system and other economic features, the police methods and tortures employed by the emperors against their ministers and subjects, relations between the emperors and eunuchs, and the fall of the dynasty which was hastened, if not brought on, by the eunuchs' abuse of power. The copious references unfortunately lack page numbers. (1,500)

2.1.11 Li Yen 李棪, Tung-lin tang chi k'ao 東林黨籍考 (A study of the members of the Tung-lin "party"; Peking: Jen-min ch'u-pan-she, 1957), 17 + 185 pp. (?)

2.1.12 Wu Han 吳晗, Chu Yuan-chang chuan 朱元璋傳 (Biography of Chu Yuan-chang; Shanghai: San-lien shu-tien, 1949), 4 + 301 pp., 6 plates.

Wu Han's biography is considered to be the most authoritative account of the life and reign of the first Ming emperor (1328-98). Written in a vivid, vernacular narrative style, it describes Chu Yuan-chang's early days as a Buddhist monk, his rise from a common soldier with the "Red Turban" rebels to become a general, the military campaigns which culminated in his ascension to the throne in 1368, and the institutional developments during the three decades of his rule. Chu's personality and character are also discussed with reference to his family relations and to what presumably were psychological maladies in his final years. The detailed biography is documented from numerous sources of and on the Ming period. An appendix gives a chronology of Chu's life. The author's postscript mentions an earlier and much shorter version of this work published in 1944 (as Ming t'ai-tsu 明太祖 and, in a second edition, as Ts'ung tseng-po tao huang-ch'üan 從僧鉢到皇权 [From the alms bowl of a monk to the power of an emperor]). The postscript also lists some twenty articles on Ming history written by the author and published between 1932 and 1943. (5,000)

2.1.13 Chu Ch'i 朱偰, Cheng Ch'eng-kung -- Ming-mo chieh-fang T'ai-wan ti min-tsu ying-hsiung 鄭成功--明末解放台湾的民族英雄 ("Koxinga"--a national hero who liberated Taiwan at the end of the Ming; Wuhan: Hu-pei jen-min ch'u-pan-she, 1956), 68 pp., 3 plates.

This biography of Cheng Ch'eng-kung (1624-62) deals with: (1) his family background in the chaotic final years of the Ming period; (2) his battles against the Ch'ing forces and campaigns to restore Ming rule; (3) his "liberation" of Taiwan from the Dutch; (4) his administration of Taiwan: land reclamation and encouragement of immigration. As the title indicates, the author's chief emphasis is on Cheng as a patriotic hero whose prime mission in life was to liberate Taiwan and the mainland of China from domination by a foreign race. The story is thus meant to inspire a present-day parallel. A bibliography is appended. (12,000)

2.1.14 Wu Tzu-chin 吳紫金 and Hung Pu-jen 洪卜仁, Cheng Ch'eng-kung shou-fu T'ai-wan chi 鄭成功收复台湾記 (An account of the recovery of Taiwan by Koxinga; Foochow: Fu chien jen-min ch'u-pan-she, 1955), 30 pp. (23,140)

2.1.15 Chu Chieh-ch'in 朱傑勤, Cheng Ch'eng-kung shou-fu T'ai-wan shih-chi 鄭成功收复台湾事蹟 (Events in Koxinga's recovery of Taiwan; Shanghai: Hsin chih-shih ch'u-pan-she, 1956), 50 pp. (35,000)

2.1.16 Chu Tung-jun 朱東潤, Chang Chü-cheng ta chuan 張居正大伝 (A biography of Chang Chü-cheng; Wuhan: Hu-pei jen-min ch'u-pan-she, 1957, originally publ. by K'ai-ming shu-tien in 1945), 27 + 391 pp.

Chang (1525-82) was a leading official and prime minister (1572-82) under the Ming emperors Mu Tsung and Shen Tsung. His own power, based on enhancing the centralizing tendencies of the late Ming government, was enormous, and earned him sharp criticism from the Censorate with whom he battled throughout his term in office. (?)

2.1.17 Li Kuang-pi 李光璧, Ming-tai yü-Wo chan-cheng 明代御倭战争 (Defensive wars of the Ming dynasty against the Japanese pirates; Shanghai: Shang-hai jen-min ch'u-pan-she, 1956), 94 pp.

This book deals with the "Japanese" piratic disturbances on the southeast China coast during more than a century between 1430 and 1565. The author describes the repeated raids on Chekiang, Fukien, and Kwangtung and the defense led by Ch'i Chi-kuang 戚繼光 and others. The economic factors that produced the Japanese pirates (actually including many renegade Chinese) and the inducement offered by the growing prosperity of the southeast Chinese coastal regions in the sixteenth century provide an introductory theme. The narrative is patriotic in emphasis, but well-documented from the primary sources. (20,000)

2.1.18 Ch'en Mao-heng 陳懋恒, Ming-tai Wo-k'ou k'ao lueh 明代倭寇考略 (A brief study of Japanese pirates in the Ming dynasty; Peking: Jen-min ch'u-pan-she, 1957), 229 pp., 1 map.

This is a reprint of Yenching Journal of Chinese Studies Monograph Series, No. 6 (1934). No mention of this fact appears in the volume. (8,700)

2.1.19 Lai Chia-tu 赖家度 and Li Kuang-pi 李光璧, Ming-ch'ao tui Wa-la ti chan-cheng 明朝对瓦剌的战争 (The Ming dynasty war against the Oirats; Shanghai: Shang-hai jen-min ch'u-pan-she, 1954), 58 pp.

A concise account of the invasion of Ming China by the Mongol federation of Oirats in the 1440's. After a brief discussion of the society and economy of the Ming dynasty, its northern border defense, the rise of the Oirat tribes and their aggressive activities, the narrative centers on the "T'u-mu-pao incident" 土木堡 when the emperor was captured in 1449, the siege of Peking, and the final Ming victory, which allegedly demonstrate both the growing decadence of the Ming rulers and the patriotic fervor of the Chinese people. Sources include the Ming-shih 明史, Ming shih-lu 明實錄 and other works of the Ming period. (21,000)

2.1.20 Yeh Hu-sheng 葉蠖生, Ming-mo nung-min ch'i-i-chün lien Ming k'ang Man hsiao-shih 明末农民起义軍联明抗满小史 (An informal history of the peasant armies' alliance with the Ming in opposition to the Manchus at the end of the Ming dynasty; Peking: Jen-min ch'u-pan-she, 1951), 37 pp. (5,000)

2.1.21 Research Institute of the Faculty of Arts and Letters, Peking University, ed., Ming-mo nung-min ch'i-i shih-liao 明末农民起义史料 (Historical materials on peasant risings at the end of the Ming dynasty), in *Ming Ch'ing shih-liao ts'ung-k'an* 明清史料叢刊 (Ming and Ch'ing historical materials series; Peking: K'ai-ming shu-tien, 1952), 37 + 529 pp., 8 plates.

The materials in this volume chiefly concern the revolt of Li Tzu-ch'eng 李自成 and peasant risings related to it, plus some items on the social background of the period. A total of 220 documents are included, all from the Main Archive of the Ch'ing Grand Secretariat (Ch'ing Nei-ko ta-k'u tang-an 清內閣大庫檔案). An interesting account of this archive and its disposition is given in the preface by Cheng T'ien-t'ing 鄭天挺. The documents, covering the years 1628-48, are arranged chronologically, with a final section containing items that cannot be precisely dated. Editorial marks have been supplied to indicate illegible or miswritten characters, omissions, deletions and the like, but the texts have not been punctuated. A full table of contents is supplied, and an appendix contains a table of major events, 1628-46. (6,000)

2.1.22 Li T'ien-yu 李天佑, *Ming-mo Chiang-yin Chia-ting jen-min ti k'ang Ch'ing tou-cheng* 明末江陰嘉定人民的抗清鬥爭 (The anti-Ch'ing struggles of the people of Chiang-yin and Chia-ting [Kiangsu] at the end of the Ming; Shanghai: Hsueh-hsi sheng-huo ch'u-pan-she, 1955), 56 pp.

An account of the fight against the Manchu forces by the people of Chiang-yin and Chia-ting prefectures in Kiangsu in 1645, based largely on half a dozen works by authors in the Ch'ing period (two of which are included in the 1911 Commercial Press edition of *T'ung shih* 痛史). It is written in a popular narrative style and is apparently intended for readers of middle school level.(?)

painful history?
通史

2.1.23 Lai Chia-tu 賴家度, Ming-tai Yün-yang nung-min ch'i-i 明代鄖陽農民起义 (Peasant risings in the Yün-yang prefecture [Hupei province] in the Ming dynasty; Wuhan: Hu-pei jen-min ch'u-pan-she, 1956), 65 pp., 1 map. (11,000)

2.1.24 Hsieh Kuo-chen 謝国楨, Nan-Ming shih-lueh 南明史略 (A brief history of the southern Ming; Shanghai: Shang-hai jen-min ch'u-pan-she, 1957), 2 + 238 pp.

A fairly popular but well-documented account of the final decade of the Ming dynasty after the Manchus had taken Peking and the Ming court had fled to Nanking. The forces that resisted the Manchus are described region by region, from Hopei to Taiwan (Koxinga). Following the current Communist interpretation of Chinese history, emphasis is laid on peasant risings and the people's support of the Ming remnants. A chronological chart of major events, 1528-1851, on both the Ming and Ch'ing sides, is given at the end of the book. The author is a professor at Nankai University in Tientsin. (12,000)

2.2 CH'ING INSTITUTIONS

Apart from 2.2.5, there is no Marxism-Maoism in the studies included in this section. Item 2.2.3 in particular should be a valuable research aid. Strictly speaking, the account of the compilation of the Ch'ing shih kao belongs to the period of the early Republic; it is included here because conditions in the Bureau of History probably changed little with the end of the Manchu Dynasty. Item 2.2.9 is included in this section out of desperation; the documents included therein, of course, go beyond institutional history, and shed light on many aspects of the late Ch'ing period. Additional material on Ch'ing institutions is included in 1.4.1, in sec. 2.1 (Ming history), in sec. 2.3, and in Section 4 (economic history).

2.2.1 Chao Ch'üan-ch'eng 趙泉澄, Ch'ing-tai ti-li yen-ko piao 清代地理沿革表 (Tables of changes in administrative areas in the Ch'ing dynasty; Peking: Chung-hua shu-chü, 1955), 14 + 204 pp., 16 tables on folded sheets.

This work first appeared in parts in the periodical Yü-kung pan-yueh-k'an 禹貢半月刊 in the 1930's. In 1941 it was published by K'ai-ming shu-tien in revised form, but copies of that edition are now extremely difficult to obtain. The present edition has a preface by Ku Chieh-kang 顧頡剛. In addition to the sixteen large tables that show the changing names, jurisdiction, etc. of the administrative subdivisions of the several provinces under the successive reign periods between 1644 and 1911, there is an accompanying text which explains the administrative organization of specific areas, and also an extensive bibliography of sources of geographic data for each province. The tables are marked in red to show, for example, the interrelationships of jurisdictions and areas under foreign jurisdiction. An index arranged by the "four-corner" system facilitates the use of the volume. (2,000)

2.2.2 Shang Yen-liu 商衍鎏, Ch'ing-tai k'o-chü k'ao-shih shu-lu 清代科舉考試述錄 (An account of the Ch'ing civil service examinations; Peking: San-lien shu-tien, 1958), 14 + 352 pp., illus.

The author, himself a chü-jen of 1904 who took more than ten examinations in a total of fifteen years, has drawn on his experience and a vast amount of research to produce an immensely detailed account of the examination system as it operated in the Ch'ing period. The text is supplemented by numerous photographs showing examination papers, examination halls, and the like. Also included are sections on the examinations for school students after the abolition of the traditional system in 1905, on the military examinations, and on the "eight-legged essay," with examples and the story of its rise and fall. Finally, a chapter on outstanding cases and legends surrounding the examinations rounds out this very comprehensive account. A list of the major printed sources consulted is given at the beginning of the book. Note that the text is written in wen-yen and not in the colloquial style. (3,000)

2.2.3 Ch'ien Shih-fu 錢實甫, Ch'ing-chi chung-yao chih-kuan nien-piao 清季重要職官年表 (Chronological tables of important official posts and their incumbents in the Ch'ing period; Peking: Chung-hua shu-chü, 1959), 9 + 270 pp.

Using data derived principally from Ch'ing shih kao 清史稿 and Ch'ing-tai cheng-hsien lei-pien 清代徵獻類編 the compiler has constructed a series of tables, covering the years 1830-1911, which show the successive incumbents year by year of the following posts: Grand Secretariat, with Assistant Grand Secretaries; Grand Council; the Six Boards; the Court of Colonial Affairs; the Censorate; governors-general; and governors. For each incumbent of each office in each year the tables contain information about assignments, promotions, honors, and the like, making this work an extremely valuable guide to official careers. In addition there are brief explanatory notes and a table of changes in the organization of the metropolitan government (offices and personnel) in the early twentieth century (pp. 113-114); and notes and a table on the establishment or removal of border officials at the end of the Ch'ing (pp. 223-224). An appendix contains: (1) a list of the persons appearing in the tables, with brief biographical data so that their careers may be followed in detail in the tables themselves; (2) a list of aliases, with accompanying better-known full names; (3) a list of posthumous titles, also accompanied by actual names; and (4) a list of provinces and hsien, accompanied by the names of the officials who orginated in each. (2,300)

2.2.4 Legal History Research Office, Bureau of Legal Affairs, State Council, comp., Ch'ing shih kao hsing-fa chih chu-chieh 清史稿刑法志註解 (Notes and commentaries on the "Essay on Justice" in Ch'ing shih kao; Peking: Fa-lü ch'u-pan-she, 1957), 122 pp.

The "Essay on Justice" in the Draft Ch'ing History is made much easier to understand for the non-expert by copious annotations (appearing as footnotes, with the text in large print above) which explain technical terms and allusions and give background information, all in pai-hua. Western dates are inserted after the Chinese dates in the text, and the pronunciation and definition of difficult words are sometimes also provided in parentheses in the text. (2,800)

2.2.5 P'eng Yü-hsin 彭雨新, Ch'ing-tai kuan-shui chih-tu 清代关税制度 (The tariff system in the Ch'ing period; Wuhan: Hu-pei jen-min ch'u-pan-she, 1956), 67 pp.

This book describes the tariff system both before and after the Opium Wars, but emphasizes the evils of the Chinese Maritime Customs under foreign administration. The text is well-documented from both Chinese and English-language sources. The conclusion sums up the detrimental effects of the "feudal" and "semi-colonial" tariff systems (before and after the Opium Wars) on the Chinese economy and polity. (8,500)

2.2.6 Yü Jung-ling 裕容齡, Ch'ing-kung so chi 清宮瑣記 (Fragmentary notes on the Ch'ing palace; Peking: Pei-ching ch'u-pan she, 1957), 85 pp., illus.

The author, daughter of the Ch'ing minister to Paris, Yü Keng 裕庚, was a lady-in-waiting to the Empress Dowager from 1903 to 1907. Her short memoir contains some fifty sections, mostly detailed descriptions or anecdotes of daily life in the Ch'ing Palace, centering around the Empress Dowager but also reflecting the institutions of Palace life and major historical events outside. There are also photographs with identifications of personnel, and a didactic postscript by Shang Tzu-kung 商子工. (72,000)

2.2.7 P'an Chi-chiung 潘際坰, Mo-tai huang-ti mi-wen 末代皇帝秘聞 (Confidential interview with the last emperor; Hong Kong: Wen-tsung ch'u-pan-she, 1957), 2 vols., 189 pp. + illus.

Though published in Hong Kong, this book was written in Peking and is the result of a ten-day interview with P'u-i 溥仪 (the Hsüan-t'ung Emperor, 1908-12) while he was in prison in Fu-shun, Manchuria. From the recollections of his childhood, one gets a few glimpses of the trivial side of Ch'ing palace life in its last days. The principal preoccupation of the book appears to be the personality and private life of a wrteched man caught up in events he could not understand, much less control. (?)

2.2.8 Chu Shih-ch'e 朱師轍, Ch'ing shih shu-wen 清史述聞 (An account of [the compilation of] the Ch'ing history; Peking: San-lien shu-tien, 1957), 14 + 432 pp.

The author was both the son of one of the "assistant compilers" (hsieh-hsiu 協修) of the Ch'ing shih kao (CSK) and later an "assistant compiler" himself. Thus he is in a unique position to shed light on many aspects of the compilation of the CSK, including editorial policies and precedents, the collection of materials, the problems of the Bureau of Ch'ing History (Ch'ing shih kuan 清史館), and the individual compilers and their contributions. These factors help to explain some of the flaws in the CSK, which was published in haste in 1928, with numerous errors and without official sanction, and then was banned by the KMT government for its unfavorable treatment of anti-dynastic movements. Although the manuscript of this volume was completed in 1955, it is history in the "old style," consisting largely of extracts from the communications sent by the compilers to the Bureau chief and the author's comments in wen-yen. (3,000)

2.2.9 Chu Shou-p'eng 朱壽朋, comp., Kuang-hsü ch'ao tung-hua lu 光緒朝東華錄 (Records of the Kuang-hsü reign; Peking: Chung-hua shu-chü, 1958, first ed. 1909), 5 vols., 20 + 6024 pp.

A reprint based on the 1909 edition, with a number of useful changes (e.g., the addition of punctuation, replacement of the original chüan divisions by the serial numbering of documents under each month, and paragraphing). The publisher's note explains briefly the background of the publication of this compilation of official documents of the period 1875-1908. It also makes the usual statement that this era saw the transformation of China into a semi-colonial, semi-feudal state, and that these documents, though entirely from the official side, are valuable as source materials on the "economic life, racial, and class contradications" of the final years of the Ch'ing dynasty. (1,200)

2.3 CH'ING HISTORY to 1800

The first two items in this section deal with Manchu institutions prior to the conquest of China. In general, the process whereby the Nu-chen (Jürched) tribes were organized into a force capable of conquering and ruling China is described by the mainland historians as the development of a feudal society out of the breakdown of Manchu tribalism and the slave system which characterized it. Items 2.3.3-7 are collections of source materials on early Ch'ing history which are arranged here in chronological order. The first of these deals with the problem of peasant uprisings, while the others are recent reprints of older sources. See also the general surveys in secs. 1.1 and 1.4, Section 4 (economic history), and secs. 5.1 and 5.5-6 (intellectual and cultural history).

2.3.1 Mo Tung-yin 莫東寅, Man-tsu shih lun ts'ung 滿族史論叢 (A collection of essays on Manchu history; Peking: Jen-min ch'u-pan she, 1958), 205 pp.

This book consists of four substantial articles on: (1) the social system of the Jürched tribes in the early Ming; (2) the breakdown of tribalism and the establishment of a Jürched state in Chien-chou 建州 in the late Ming, which the author sees as the consequence of development from a slave towards a feudal type of society; (3) the Eight Banner System of the Ch'ing just before the conquest of China (described as proto-feudal on the way to becoming feudal); and (4) Manchu shamanism in the early Ch'ing. Proper acknowledgment is made to the Marxist classics for the author's theoretical views, but the copious footnotes refer mainly to Ming and Ch'ing sources, the Korean Shih-lu, and modern studies. (2,300)

2.3.2 Wang Chung-han 王鍾翰, Ch'ing shih tsa k'ao 清史雜考 (Miscellaneous studies on Ch'ing history; Peking: Jen-min ch'u-pan-she, 1957), 323 pp.

This volume contains eight scholarly articles, some of which were previously published in the Yen-ching hsueh-pao 燕京学报 or other periodicals. Four were written before 1949 in the literary style: "The So-lun 索倫 tribal origin of the Ta-hu-erh 達呼爾," "The Eight Mongolian Banners of early Ch'ing," "The usurpation of Shih-tsung 世宗 (Yung-cheng 雍正)," and "The compilation of the Ch'ing [supplements] to the San t'ung 三通." Two articles were written after 1949 and in pai-hua: "The social and economic condition of the Manchus in the Nurhaci period," and "The process of feudalization of the Manchus under Abahai 皇太極." The author claims (correctly, it would seem) that these are "experimental writings" of a beginner in Marxism-Leninism. Two appendices discuss the Grand Council (these are notes from a lecture given by Professor Teng Chih-ch'eng 鄧之誠 at Yenching in 1937) and the Tsungli Yamen (using data culled from the I-wu shih-mo 夷務始末 documents). (9,600)

2.3.3 Hsieh Kuo-chen 謝国楨, ed., Ch'ing-ch'u nung-min ch'i-i tzu-liao chi-lu 清初農民起义資料輯錄 (Collected records of peasant risings in the early Ch'ing dynasty; Shanghai: Hsin chih-shih ch'u-pan-she, 1956), 5 + 404 pp.

This volume begins with a general account of peasant revolts in various regions in China between 1644 and 1721 and some brief generalizations about the character of these "anti-feudal" and "anti-Ch'ing" uprisings (pp. 1-49). The remainder of the text consists of passages concerning peasant rebellions extracted chiefly from official sources such as the Ch'ing shih-lu 清實錄, Ch'ing shih kao 清史稿, Ming Ch'ing shih liao 明清史料, and Wen-hsien ts'ung-pien 文献叢編. These are augmented with extracts from local gazetteers and private writings of various sorts. The original sources are indicated, and the editor has added occasional explanatory notes. The greater part of the materials deals with activities of the "peasant armies" in 1644-73 and is arranged by region: Hopei-Shantung-Honan, Kiangsu-Anhwei-Chekiang, Hupei-Hunan (including Li Tzu-ch'eng's followers), Yunnan-Kweichow, Shansi-Shensi-Kansu, and Kwangsi-Fukien-Kwangtung. The years 1673-1721 are covered in one final section. A "chronological table of early-Ch'ing peasant risings" is appended (pp. 391-404). (6,000)

2.3.4 Hsiao Shih 蕭奭, Yung hsien lu 永憲錄 (A record of "eternal principles"), Ch'ing-tai shih-liao pi-chi ts'ung-k'an 清代史料筆記叢刊 (Ch'ing historical materials and notes series; Peking: Chung-hua shu-chü, 1959), 7 + 424 pp.

This is a compilation dealing with political events during the years 1722-28, the materials for which were gathered mainly from court reports, edicts, and memorials. It is written in the style of informal history, by an author about whom nothing is known. This reprint is based on an edition in the possession of the historian Teng Chih-ch'eng, who has written a colophon dated 1955. (1,900)

2.3.5 Juan K'uei-sheng 阮葵生, Ch'a yü k'o-hua 茶餘客話 (Chit-chat after tea), in Ming Ch'ing pi-chi ts'ung-k'an 明清筆記叢刊 (Ming and Ch'ing notes series; Shanghai: Chung-hua shu-chü, 1959), 2 vols., 27 + 728 pp.

The present version of this work (written in 1771) is based largely on an 1888 edition, with some sections added from an earlier edition. The numerous passages in a total of twenty-two chüan are given titles and listed in the table of contents, enabling the reader to scan the material, which ranges in subject from politics, geography, intellectual history, and literature, to food and tea-drinking. The passages concerning the Grand Secretariat (where the author once worked in the document-copying section) and the examination system (which made him a chin-shih) are more informative than his cursory comments on various trivia. (2,900)

2.3.6 Feng Ch'eng-chün 馮承鈞, Hai-lu chu 海錄註 (Notes on Hai lu ["Overseas record," by Hsieh Ch'ing-kao 謝清高]; Shang-hai: Commercial Press, 1955), 14 + 83 pp.

A reprint of a 1937 Commercial Press publication consisting of annotations on Hai lu, an account of foreign lands in Asia and Europe visited by Hsieh Ch'ing-kao (1765-1821) between the years ca. 1782 and 1795. The annotations identify the place names in the original text by their modern names, in Chinese and in English, and elucidate other terms and allusions in the text. (2,500)

2.3.7 Yü Cheng-hsieh 俞正燮, K'uei-ssu ts'un-kao 癸巳存稿 (Remaining manuscripts compiled in the year K'uei-ssu [1833]; Shanghai: Commercial Press, 1957), 27 + 496 pp.

This book was first published in 1847 in Lien yün i ts'ung-shu 連筠簃叢書, with other editions subsequently. The present version is a corrected and augmented reprint of the 1937 Commercial Press edition. As the title implies, this work consists of material left out of another collection of Yü's writings, K'uei-ssu lei-kao 癸巳類稿 (republished 1957, but not seen by us). The fifteen chüan of this book contain a great number of items of varying lengths on many subjects, including the classics, history, geography, and social customs.

Perhaps these two works have been reprinted because of Yü's relatively liberal ideas, in particular his defense of the rights of women and attacks on the prevailing standards of morality. (3,500)

2.4 THE OPIUM WAR

The Opium War, described as the beginning of foreign aggression against China, is the starting place for modern history in mainland China as it is for all Chinese nationalists, Communist and non-Communist alike. The particular Communist twist in this now standard interpretation is to describe Chinese society prior to the Opium War as "feudal" and to assert that in the Opium War and its aftermath Chinese feudal society was transformed into a "semi-feudal, semi-colonial" society which lasted until the establishment of the People's Republic in October 1949. But to accept this historical periodization apparently can raise some difficult questions for the Chinese historian who is both a Marxist and an emotional nationalist. In the following remarks by a contemporary historian, Li Shu, in connection with the current discussions of historical periodization in Communist China, such a question for a moment rises to the surface and then is drowned in sophistry:

> There remains one important question. If we employ a foreign war of aggression against China to mark a division point in the periodization of our history, do we not then become proponents of external causation? My answer is to deny this. Comrade Mao Tse-tung has pointed out that the correct use of [historical] materialism does not at all deny a role to external causative factors; but external causes only manifest themselves through internal causes. Foreign capitalist aggression against China has influenced China to undergo internal changes. This means that Chinese society internally already

> possessed the prerequisite for the appearance of change; and this prerequisite was the high degree of development that China had obtained during the long period of feudal society. At the time of the encroachment of foreign capitalism, if China had been no more than a primitive tribal society, the appearance of a bourgeoisie and a proletariat would have been impossible, no individuals with a developed consciousness could have emerged at all, and there could not have been a conscious revolutionary movement....Comrade Mao Tse-tung has pointed out that China's revolution "is not merely a formless uprising produced by the incursion of Western thought. It is a war of resistance provoked by imperialist aggression." This resistance is a conscious resistance which developed steadily in the period after the Opium War. Therefore, it is reasonable for us to take the Opium War as a division point to mark the great revolutionary period that followed it. (Li Shu, "Chung-kuo ti chin-tai shih yü ho shih?" [When was the beginning of modern history in China?], Li-shih yen-chiu [Historical studies], 1:1-16 [1958]).

There is an obvious anxiety in this statement that to assign so large a role to foreign incursions in the structuring of China's modern history verges on abandoning the belief that this history was capable of autonomous development. This is almost to cast doubt on the value of one's own past--no small emotional wrench for a hyper-nationalist Chinese. An uneasy resolution is reached by asserting that "external causes only manifest themselves through internal causes" and that "Chinese society internally already possessed the prerequisite for the appearance of change." Perhaps it is comforting to dispose of the problem in this manner, but the inherent instability of the answer is revealed in the oscillations of Chinese Communist historiography in the past few years, such as is discussed below in sec. 4.3 in connection with the incipient capitalism controversy.

Item 2.4.2 gives a good cross-section of the types of material on the Opium War which have appeared in the past ten years in Communist China. Generally speaking, it is in the collection and publication of source materials (such as 2.4.3-7) that the most impressive achievements both as regards quality and quantity, are to be found in the field of modern history. See also the survey treatments in sec. 1.4.

2.4.1 Pao Cheng-ku 鮑正鵠, Ya-p'ien chan-cheng 鴉片戰爭 (The Opium War; Shanghai: Hsin chih-shih ch'u-pan-she, 1954), 140 pp.

Presumably written for the general public, this volume is a simplified and often cliché-ridden account of the Opium War, its causes and consequences. It contains three useful maps. (35,000)

2.4.2 Lieh Tao 列島, ed., Ya-p'ien chan-cheng shih lun-wen chuan-chi 鴉片战争史論文專集 (Special collection of essays on the Opium War; Peking: San-lien shu-tien, 1958), 2 + 380 pp.

This collection consists of twenty-four articles out of a total of over eighty on the subject of the Opium War which appeared in mainland Chinese newspapers and periodicals between 1949 and the end of 1957. (The titles of the articles not included here are listed in an appendix, pp. 374-380, with the names and dates of the publications in which they first appeared.) The twenty-four articles are for the most part written by well-known historians, including such senior professors as Ch'i Ssu-ho and Fung Yu-lan 馮友蘭. Among the topics are the British and American consulates in China and opium merchants, the voyage of the British ship "Amherst," and local resistance to the British in Kwangtung. Some of the pieces are not directly concerned with the Opium War but deal with events in the same period (e.g., the American coolie trade, or American missionaries in

China before 1858). Some are quite short (5-10 pp.). The longest one (60 pp.), on the relations between the development of the British textile industry from the 1830's to the 1850's and the two Opium Wars, is by the economic historian Yen Chung-p'ing . (3,000)

2.4.3 Ch'i Ssu-ho 齊思和 , Lin Shu-hui 林樹惠 , Shou Chi-yü 壽紀瑜 , eds., Ya-p'ien chan-cheng 鴉片戰爭 (The Opium War), Modern Chinese Historical Materials Series, No. 1 (Shanghai: Shen-chou kuo-kuang she, 1954), 6 vols., 3,757 pp.

Source materials covering the period from the mid-1830s to the conclusion of the Treaty of Nanking in August 1842, arranged chronologically under six topics: (1) British economic encroachment prior to the Opium War; (2) the beginning of the opium prohibition movement; (3) the struggles against the British and the opium trade under the leadership of Lin Tse-hsü; (4) British military aggression and the resistance of the Chinese people; (5) the Treaty of Nanking and the reaction of the Chinese people to the treaty settlement; and (6) accounts of the war as a whole. The bulk of these materials consists of edicts and memorials, letters and diaries of contemporaries, and essays, poems and other writings of eye-witnesses both official and private. It has been estimated by Hsin-pao Chang in The Journal of Asian Studies, 17:64 (Nov. 1957), that included in the main body of these six volumes are 955 pages of manuscripts published now for the first time, 1,216 pages of materials from rare or little-known Chinese books, 491 pages of translated materials, and only 865 pages of source materials well-known to specialists. The editors have supplied several useful appendices (e.g., biographies of the leading Chinese and English figures in the conflict, and lists of Chinese and English officials). Finally, there is an extremely valuable annotated bibliography (144 pp.) prepared by Ch'i Ssu-ho. A second edition, with minor revisions, was published in 1957 by the Shanghai jen-min ch'u-pan she. (8,000)

2.4.4 Ch'i Ssu-ho 齊思和, ed., Huang Chueh-tzu tsou-shu Hsü Nai-chi tsou-i ho-k'an 黃爵滋奏疏許乃濟奏議合刊 (The memorials of Huang Chüeh-tzu and Hsu Nai-chi; Peking: Chung-hua shu-chü, 1959), 221 pp.

Eighty-six memorials of Huang Chüeh-tzu (1793-1853), from a manuscript entitled "Huang shao-ssu-k'ou tsou shu" 黃少司寇奏疏 kept at the Peking Museum, comprise Part I (pp. 1-195) of this volume. Except for eleven items which were included in slightly abridged form in the Tao-kuang I-wu shih-mo 夷務始末, these memorials have never before been published. Dealing with such topics as administration and military defense during the Opium War period, the memorials are arranged in twenty chüan according to the office held by Huang when he wrote them. The memorials of Hsü Nai-chi (chin-shih of 1809), numbering thirteen (pp. 197-221), are based on a manuscript entitled "Hsü T'ai-ch'ang tsou-i" 許太常奏議 kept at the Nanking Library. Among these is Hsü's controversial proposal of 1836 to relax the opium prohibition, which is also included in the Tao-kuang I-wu shih-mo. (1,900)

2.4.5 Liang T'ing-nan 梁廷枏, I fen wen chi 夷氛聞記 (An account of the barbarians' ill wind), in Chin-tai shih-liao pi-chi ts'ung-kan 近代史料筆記叢刊 (Collection of modern historical materials and notes; Peking: Chung-hua shu-chü, 1959), 10 + 172 pp.

This work, originally published ca. 1874, is reprinted as an important primary source on the Opium War. (It is also included in 2.4.3). The author is praised for his recognition of the British aggressive attitude and of the indomitable spirit of the Chinese people. But he was not, in the words of Shao Hsün-cheng 邵循正, who is the collator and annotator of this edition, free from the "branding of his time and class," i.e., he exaggerated the role of the gentry, underestimated the force of the masses, and mistook reactionary bureaucrats for heroes. Mr. Shao compares several popular editions of the book with the original edition and notes the discrepancies in parenthetical remarks in the text. He also supplies the English for names transliterated into Chinese. (4,100)

2.4.6 Shanghai branch of the Institute of Historical Research, Chinese Academy of Sciences, comp., Ya-p'ien chan-cheng mo-ch'i Ying-chün tsai Ch'ang-chiang hsia-yu ti ch'in-lueh tsui-hsing 鴉片战争末期英軍在長江下游的侵略罪行 (The criminal aggression of English troops in the Lower Yangtze region in the last phase of the Opium War; Shanghai: Shang-hai jen-min ch'u-pan-she, 1958), 2 + 392 pp., plates.

The bulk of this volume (pp. 31-252) consists of selections translated into Chinese from five contemporary English accounts of the Opium War: W. D. Bernard, Narrative of the Voyages and Services of the Nemesis from 1840 to 1843 (London, 1844); Granville G. Loch, The Closing Events of the Campaign in China: The Operations in the Yang-tze-kiang and Treaty of Nanking (London, 1843); John Ouchterlony, The Chinese War (London, 1844); Alexander Murray, Doings in China (London, 1843); and A. Cunynghame, Opium War: An Aide de Camp's Recollections of Service in China (London, 1844). There are, in addition, documents selected from the Tao-kuang I-wu shih-mo, which are intended to demonstrate the defeatist Manchu policy "which made the loss of the Opium War inevitable" (pp. 253-309); materials illustrating the resistance offered by local officials and soldiers to the British attackers (pp. 311-342); and materials illustrating popular resistance to the British and the Ch'ing government's undermining of this resistance (pp. 343-387). A number of the items under these last two heads are printed from manuscript sources. A table of transliterations of foreign names and their English equivalents is appended. (2,500)

2.4.7 A Ying 阿英, ed., Ya-p'ien chan-cheng wen-hsueh chi 鴉片战争文学集 (A literary collection on the Opium War; Peking: Ku-chi ch'u-pan-she, 1957), 2 vols., 55 + 1010 pp.

This is the first of a series of similar collections edited by A Ying (pseudonym of Ch'ien Hsing-ts'un 錢杏邨), who has been active in literary circles in Communist China. They bring together contemporary literary output purporting to show Chinese patriotism in the face of foreign aggression (see 2.8.7, 2.9.4, and 2.9.11). The editor has contributed a long introduction in which he discusses the Opium War and the "victory of the Chinese people," and the various forms of literature that demonstrate their indomitable spirit. These materials are grouped under four headings: poems and songs, novels, drama, and essays. The authors (whom the editor does not identify) range from well-known literary and political figures of that period to obscure or anonymous writers. As with the other collections, these volumes are of interest as historical materials, rather than for their literary merit. (10,000)

2.4.8 Yao Wei-yuan 姚薇元, Ya-p'ien chan-cheng shih shih k'ao 鴉片战争史實考 (An authentication of the historical facts of the Opium War); alternate title: Wei Yuan "Yang-sou cheng-fu chi" k'ao-ting 魏源"洋艘征撫記"考訂 (A textual study of Wei Yuan's "Record of Pacification of Foreign Ships"; Shanghai: Hsin chih-shih ch'u-pan-she, 1955, 6 + 164 pp.

An earlier version of this book appeared in Kweichou in 1943, with an informative preface by T. F. Tsiang (Chiang T'ing-fu 蔣廷黻) and another by Kuo T'ing-i 郭廷以. These prefaces are neither included nor mentioned in the present edition; the author's own foreword has been revised to include the now standard labeling of the Opium War as the beginning of the semi-colonial semi-feudal period of Chinese society. The text, which is a commentary on Wei Yuan's account of the Opium War (in Sheng wu chi 聖武記, 1846 ed.) is also somewhat revised but remains in the literary style. It compares Wei Yuan's writing on this subject with various other works and documents, Chinese and English, and so reveals inaccuracies in Wei's version and offers a fuller and more rounded account. About 100 topics referred to by Wei Yuan are here cited and annotated. The present edition has a more attractive format, including a table of contents, and is more legible. (6,100)

2.4.9 Wei Chien-yu 魏建猷, Ti-erh-tz'u ya-p'ien chan-cheng 第二次鴉片战争 (The second Opium War; Shanghai: Shang-hai jen-min ch'u-pan-she, 1957), 2 + 139 pp.

A relatively detailed account of the second Opium War (1856-60), drawing heavily on the I-wu shih-mo documents. The interpretation which links Western aggression and Ch'ing political bankruptcy is standard for the People's Republic of China. It would be of interest to compare this treatment section by section with the relevant parts of H. B. Morse's The International

Taiping Rebellion

Relations of the Chinese Empire, Vol. I. (39,000)

2.5 THE TAIPING REBELLION

If the Opium War is the beginning of China's transformation into a "semi-colonial, semi-feudal" status, the defeat of the Taiping Rebellion (1851-64) marks the watershed, after which the decay of imperial China and its enslavement by the "foreign imperialist aggressors" proceeds ever more rapidly. The Taiping movement is currently interpreted in mainland China not only as a peasant revolution against the feudal Manchu regime, but also as a struggle of the people against the foreign aggressors who assisted the Manchu ruling class in suppressing the revolutionary movement. In the new historical tradition which they have sought to construct, Chinese Communist historians see the Taiping Kingdom as a direct ancestor of the Chinese Communist revolutionary movement. The dynamism of both can be traced in large part to peasant unrest. What made the modern movement successful while its predecessor failed, say the mainland historians, was the new element of proletarian leadership provided by the Communist Party under the leadership of Mao Tse-tung. The Taipings failed because they lacked a disciplined organization and a guiding ideology, but their contribution to the Chinese revolutionary tradition is seen as comparable to that of the Revolution of 1789 in France.

This is thus one of the most popular subjects with the mainland historians. The first four items in this section (2.5.1-4) are general histories of the Taiping movement, including two versions of a survey by Lo Erh-kang, the most important single scholar of the Taipings. Items 2.5.5-10 are collections of source materials, again a matter to which a great deal of effort is being devoted. These are followed by studies of individual Taiping leaders (2.5.11-17). The remaining titles are monographic treatments of aspects of the Taiping movement, including many important studies by Lo Erh-kang. Professor Lo's scholarly standards seem to have remained intact, but in his prefatory material he now makes the expected bow to Marxism-Leninism-Maoism. See also the surveys discussed in sec. 1.4., sec. 2.6, items 2.7.2-3, and 6.1.16 (a comprehensive bibliography of the Taiping Rebellion).

2.5.1 Lo Erh-kang 羅爾綱, T'ai-p'ing t'ien-kuo shih-kao 太平天国史稿 (A draft history of the Taiping Kingdom; Peking: K'ai-ming shu-tien, 1951), 10 + 285 pp.

This "draft history" is written in the literary style and uses the classical historical format: I, annals; II, tables; III, treaties; and IV, biographies. Part I deals with the Heavenly King, Hung Hsiu-ch'üan 洪秀全 and his son, the young Heavenly King. Part II lists the anti-Ch'ing battles of the Nien Army, the uprisings of secret societies before and during the Taiping period, the kings and princes of the Taiping Kingdom, and foreign generals. Part III includes "monographs" on Taiping religion, society, calendar, army, bureaucracy, ceremonies, civil examinations, and official documents. Part IV includes over sixty biographies of Taiping leaders. A fairly comprehensive bibliography, including some contemporary English publications, is appended (pp. 274-285). (10,000)

2.5.2 Lo Erh-kang 羅爾綱, T'ai-p'ing t'ien-kuo shih kao 太平天国史稿 (A draft history of the Taiping Kingdom; Peking: Chung-hua shu-chü, 1955), 12 + 393 pp., 15 plates; enlarged ed., 1957, 15 + 438 pp., 18 plates.

A revised version of 2.5.1, this work retains substantially the same material and the same organization, but it also shows these changes: (1) The wen-yen style has been somewhat modified, and many standard literary expressions are now replaced with Marxist phraseology. (2) Less attention is given to the Taiping religion and more to Taiping society. (3) Sections on foreign relations, art, and the economy have been added. (The author states that the first edition was in error in overlooking the accomplishment of the Taiping Kingdom in "resisting foreign capitalist aggression.") (4) New biographies have been added, including those of Triad and Nien leaders. (5) The bibliography in the first edition is omitted. Professor Lo has written a fairly lengthy preface which defends his use of the classical historical format. It also describes the Taiping uprising as a peasant revolution without proletarian leadership and with inherent feudal-progressive contradictions. An enlarged edition of this work was issued in 1957 by the same publisher. It has an additional author's preface, which says that the 1955 edition may have overstated the "progressivism" of the Taipings and understated their backwardness. In a new introductory section (33 pp.--the manuscript was read before the National Committee of the Chinese People's Political Consultative Conference in August 1956) Professor Lo addresses himself to such questions as the periodization of the Taiping movement and the nature of the Nien Army. He holds that the Nien Army was a subsidiary of the Taipings which took part in regional uprisings but lacked well-defined objectives and a strong organization of its own. (18,000)

2.5.3 Hua Kang 华刚, T'ai-p'ing t'ien-kuo ko-ming chan-cheng shih 太平天国革命战争史 (A history of the Taiping revolutionary war), rev. ed. (Shanghai: Shang-hai jen-min ch'u-pan-she, 1955), 212 pp., 1 map.

This volume, the author states, was first published in 1949 (in haste and in "imperfect" form) to counteract the "distorted and slanderous" accounts of the Taiping movement circulated by "bourgeois historians." Between 1949 and 1952 the book went through seven printings, and in 1952 a Russian translation was published in Moscow. It is a relatively detailed account of the Taiping movement from the Chin-t'ien uprising in 1850 through the final encounters with the Ch'ing forces in 1863-64 (with the Nien and other uprisings as a coda). Hua Kang's conclusion sees the Chinese People's Republic as a realization of the promise of the Taiping revolution, but the bulk of the text is narrative, with occasional explicit interpretation. See also a review by Hsiao Kung-ch'üan in Far Eastern Quarterly, 12.2:218-220 (Feb. 1953); and another by Teng Ssu-yü in ibid., 12.3:319 (May 1953). (13,000)

2.5.4 Mou An-shih 牟安世, T'ai-p'ing t'ien-kuo 太平天国 (The Taiping Kingdom; Shanghai: Shang-hai jen-min ch'u-pan-she, 1959), 6 + 464 pp., map, illus.

Possibly designed as a college textbook, this volume combines the use of many of the major primary and secondary sources on the Taiping movement with selected works of Marx, Lenin, and Mao. The history of the Taipings is divided into seven stages, beginning with the period of "revolutionary ferment" (1843-51), through several phases of victory or difficulty, and ending with "continued revolutionary struggles" after the fall of Nanking (1864-68). The sixth stage, "period of direct fighting with foreign aggressors to protect the revolutionary cause (1861-64)," reflects the current interest among mainland historians in stressing the Taiping conduct of foreign relations. For his conclusion, the author borrows a remark of Marx on the French revolution as being directly applicable to the Taiping movement (what was destroyed was not the revolution but vestiges of the pre-revolutionary tradition). Three appendices give a table of major events from 1843 to 1868, a glossary of Western names in the text, and a bibliography of reference works. (11,000)

2.5.5 Hsiang Ta 向達 et al., eds., T'ai-p'ing t'ien-kuo 太平天国 (The Taiping Kingdom), Modern Chinese Historical Materials Series, No. 2 (Shanghai: Shen-chou kuo-kuang she, 1952), 8 vols., 3405 pp.

These materials on the Taipings are grouped by the provenance of documents contained in each category. Part I contains thirty-eight official Taiping publications and more than 100 other documents by the rebels themselves (e.g., proclamations, the confessions of captured Taiping leaders, and examples of administrative records). Part II contains non-Taiping contemporary Chinese writers, including a number of rare accounts of the rebellion. Part III includes translations of a number of Western-language contemporary accounts. These for the most part were made by the non-Communist historian Chien Yu-wen 簡又文 , who is given credit, and have been published previously. Part IV includes two groups of previously unpublished military accounts of the early stages of the uprising--the memorials of Hsiang Jung 向榮 (d.1856) and the correspondence of the Manchu official Wu-lan-t'ai 烏蘭泰 (d.1852). The main strength of this collection is the relatively comprehensive assembly of source material in one publication, including some selections hitherto rare and hard to obtain. The non-Taiping material is very selective. No bibliography of sources is included, but this gap has since been filled by item 6.1.16. A new edition, which corrects typographical and other minor errors, was published in 1957 by the Shanghai jen-min ch'u-pan-she. See The Journal of Asian Studies, 17.1:67-76 (Nov. 1957), for a detailed review of this collection . (10,000)

2.5.6 Wang Ch'ung-wu 王崇武, Li Shih-ch'ing 黎世清, trans., T'ai-p'ing t'ien-kuo shih-liao i ts'ung 太平天国史料譯叢 (A collection of translated historical materials on the Taiping Kingdom; Shanghai: Shen-chou kuo-kuang she, 1954), 5 + 256 pp., 14 plates.

Part I contains material on the Taiping rebellion in the British Public Record Office, mostly consular reports from Ningpo and Shanghai during 1862-63. Part II contains twelve letters from the Shanghai branch of Jardine Matheson to the Hong Kong office (1853-62), which are from the Company's archives. Part III is an article in the London Times (Oct. 3, 1860) by Joseph Edkins of the London Missionary Society, on his interview with Li Hsiu-ch'eng. Part IV consists of six chapters from Gordon in China and the Soudan by A. E. Hake (1896). Each of the two translators has written an introduction to the material for which he is responsible, giving some background information and stressing the anti-Taiping position of the British. Mr. Li, the translator of Hake, rates this book as relatively authentic, but his view of the Gordon diary and papers is wholly unfavorable. (5,000)

2.5.7 Shensi Provincial Museum, ed., T'ai-p'ing-chün Han-chung chan-cheng shih-shih chieh-ch'ao 太平軍漢中战爭事實節鈔 (A manuscript concerning the Taiping Army's fighting in the Han River valley; Sian: Shensi jen-min ch'u-pan-she, 1957), 3 + 78 pp.

This volume results from the discovery in 1956 of a damaged manuscript in five volumes of the gazetteer of Feng-hsien, Shensi (Hsin-hsiu Feng-hsien-chih ch'u-kao 新修鳳縣志初稿 [Newly revised gazetteer of Feng-hsien, first draft]; the 1892 edition of the Feng-hsien gazetteer refers to this as the Kuo manuscript). Two of these volumes contain an account of the fighting between the Ch'ing and Taiping forces at Feng-hsien in the period 1862-66, and also some incidental material on the Muslim rebellion. These two volumes are published in edited form, with certain materials such as the militia regulations and chart of fortresses omitted. What remain are day-to-day entries on military operations. (1,700)

2.5.8 Ching Wu 靜吾 , Chung Ting 仲丁 , eds., Wu Hsü tang-an chung ti T'ai-p'ing t'ien-kuo shih-liao hsüan-chi 吳煦檔案中的太平天国史料選輯 (Selected historical materials from the archive of Wu Hsü on the Taiping Kingdom; Peking: San-lien shu tien, 1958), 6 + 300 pp.

Hitherto unknown manuscript materials of the Taiping period were discovered in Hangchow in 1953. They had been preserved by the descendants of a Ch'ing official, Wu Hsü, and consisted of documents and letters written to and by Wu while he was military-administrative taotai (ping-pei tao 兵備道) of the Suchow-Sungkiang-Taitsang district, together with Taiping documents he had collected, and his account books and rosters. The materials published in the present volume include only that part of the archive which the editors consider to be related to the "joint Sino-foreign anti-revolutionary forces" during the Taiping movement. The volume is divided into six sections: (1) documents issued by the Taipings (including two letters from Li Hsiu-ch'eng to foreign officials); (2) papers relating to the "Small Knife Society" (Hsiao-tao hui 小刀会); (3) memorials and other documents concerning relations between the Ch'ing court and Western powers; (4) documents relating to the "Ever Victorious Army" and other foreign military assistance to the government forces; (5) letters from foreign officials to Wu Hsü; (6) selected dispatches of the Hui-fang chü 会防局 ("Joint Defense Bureau," led by Shanghai gentry and local officials, which cooperated with the British and French authorities in the defense of Shanghai against the Taiping forces) between 1862 and 1864. (3,700)

2.5.9 Chin Yü-fu 金毓黻 and T'ien Yü-ch'ing 田餘慶, eds., *T'ai-p'ing t'ien-kuo shih-liao* 太平天国史料 (Historical materials on the Taiping Kingdom), second ed. (Peking: Chung-hua shu-chü, 1959), 6 + 519 pp.

The second edition of this source book differs from the first (1950) printing chiefly in that the texts have now been punctuated. As before, the editors offer no introductory remarks to their selection of documents. These materials are duplicated in part by item 2.5.5, but they include some important sources not available in the larger collection. The volume consists of four parts: (1) three official Taiping publications, one of which is by Hung Jen-kan 洪仁玕; these have appeared in installments in the journal *I-ching* 逸經; (2) seventy-five Taiping documents, including edicts, communications with foreign envoys, and rosters of troops; some of these have been published previously in the *Kuo-wen chou-pao* 国聞週報; (3) 182 documents from the Manchu side, including edicts, selections from the Gordon papers in the British Museum, and some intelligence reports; these have not been previously published; (4) miscellaneous contemporary Chinese and foreign records. Most of the materials in the last three do not appear in the *T'ai-p'ing t'ien-kuo* collection, and were copied by Hsiang Ta 向达 in the British Museum, or photographed by Wang Chung-min 王重民 in the Cambridge University Library. (3,000)

2.5.10 Exhibition Commemorating the Hundredth Anniversary of the Taiping Uprising, ed., *T'ai-p'ing t'ien-kuo ko-ming wen wu t'u-lu* 太平天国革命文物圖錄 (A pictorial volume of documents and objects of the Taiping revolution; Shanghai: Shang-hai ch'u-pan kung-ssu, 1952-55), 3 vols. Vol. 1 (*cheng-pien* 正編), 1952, 88 plates (folio); Vol. 2 (*hsü-pien* 續編), 1953, 67 plates; Supplementary Volume (*pu-pien* 補編), ed. by Kuo Jo-yü 郭若愚, Shanghai: Ch'ün-lien ch'u-pan-she, 1955, 80 plates. (8,000)

2.5.11 Shu Shih-cheng 束世澂, *Hung Hsiu-ch'üan* 洪秀全 (Shanghai: Hsin chih-shih ch'u-pan-she, 1955), 119 pp. (36,000)

2.5.12 Li Ch'un 酈純, *Hung Jen-kan* 洪仁玕 (A biography of Hung Jen-kan; Shanghai: Shang-hai jen-min ch'u-pan-she, 1957), 66 pp. (20,000)

2.5.13 Lo Erh-kang 羅爾綱, *Chung-wang Li Hsiu-ch'eng chuan* 忠王李秀成傳 (A biography of the Loyal Prince, Li Hsiu-ch'eng; Nanking: Chiang-su jen-min ch'u-pan-she, 1954), 5 + 60 pp., plates.

This sympathetic biography of Li Hsiu-ch'eng (1823-64) is written for popular consumption. It traces briefly Li's rise from a poor peasant boy to the leadership of the armies of the Taiping Kingdom, his military campaigns, and martyrdom. The material for this book is derived from Lo's scholarly works on Li Hsiu-ch'eng's biography. (12,000)

2.5.14 Lo Erh-kang 羅爾綱, *Chung-wang Li Hsiu-ch'eng tzu-chuan yuan-kao chien cheng* 忠王李秀成自傳原稿箋證

(A commentary and textual study on the original manuscript of the autobiography of the Loyal Prince, Li Hsiu-ch'eng; Peking: Chung-hua shu-chü, 1954), 210 pp., plates.

This is a revised version of *Chung-wang Li Hsiu-ch'eng tzu-chuan* (1950). Part I, "the textual study," deals with the three main editions of the autobiography of Li Hsiu-ch'eng and considers the evidence which the author believes supports his hypothesis that the manuscript was actually written by Li but was tampered with by Tseng Kuo-fan. Part II, the "commentary," consists of annotations which elucidate dates, events, and persons which Professor Lo believes to be inaccurate or unclear. As an appendix, Professor Lo offers a document entitled "Chung-wang tzu-chuan pieh-lu" 忠王自傳別錄 (An addition record pertaining to the autobiography of the Loyal Prince). This is a manuscript record of the interrogation of Li Hsiu-ch'eng by Tseng Kuo-fan and two other Ch'ing officials which first came to light in 1937. (8,000)

2.5.15 Shen Ch'i-wei 沈起煒, *Li Hsiu-ch'eng* 李秀成 (A biography of Li Hsiu-ch'eng; Peking: Chung-kuo ch'ing-nien ch'u-pan-she, 1955), 91 pp. (25,000)

2.5.16 Lo Erh-kang 羅爾綱, *Chung-wang tzu-chuan yuan-kao k'ao-cheng yü lun k'ao-chü* 忠王自傳原稿考証与論考據 (Authentication of the original manuscript of the autobiography of the Loyal Prince [Li Hsiu-ch'eng] and textual research; Peking: K'o hsueh ch'u-pan-she, 1958), 94 pp., illus.

Three articles by the foremost specialist on the Taiping Rebellion. From the first article, one gathers that he is stimulated by the new political climate to make an appraisal of the relative merits of what he calls the "old" and "new" schools of textual research. In the second article, Professor Lo defends his earlier thesis (published in 1951) that the manuscript of Li Hsiu-ch'eng's autobiography was genuine, although parts of the contents had been deleted by Tseng Kuo-fan. In the third article, he reiterates this position by illustrating the uses of verification by analysis of handwriting. A number of plates reproducing parts of the manuscript of the Li autobiography are included. (4,505)

2.5.17 Liang Hu-lu 梁岵廬, ed., Chung-wang Li Hsiu-ch'eng tzu-shu shou-kao 忠王李秀成自述手稿 (The handwritten manuscript of the autobiography of Li Hsiu-ch'eng; Peking: K'o-hsueh ch'u-pan-she, 1958), 49 pp., 16 plates (folio). (2,055)

2.5.18 Lo Erh-kang 羅爾綱, T'ai-p'ing t'ien-kuo shih chi-tsai ting-miu chi 太平天国史記載訂謬集 (Commentaries on falsehoods in the historical record of the Taiping Kingdom), Collected Essays on the History of the Taiping Kingdom, Vol. 1 (Peking: San-lien shu-tien, 1955), 170 pp.

While the essays in this volume were mostly written prior to 1949, when Professor Lo's approach to Taiping history and the authenticity of its record was that of a meticulous scholar, the preface to the present publication (the first of seven volumes of his collected essays on the Taipings) contains an added element of political ideology. He states that his former antiquarian method of textual study did not enable him to deal with overall and complex problems, and that in order to penetrate and unearth the content of history it is necessary to have a firm grasp of the Marxist-Leninist "viewpoint and position." Although, Professor Lo admits, he is "far from possessing this firm grasp," he does incorporate some Communist terminology into his hitherto apolitical writings, as the title of the first article illustrates: "A study of the falsehood of the Manchu ruling class's slanderous charge that the Taiping Army committed homicide, arson, rape, and plunder." Two articles in this volume, on Ch'ien Chiang 錢江, and the genealogy of Yang Hsiu-ch'ing 楊秀清, are substantially unchanged from versions published in 1950. Another article which also appeared in 1950, on Tseng Kuo-fan's memorial on the sack of Nanking, is now revised and somewhat expanded (in the present version Nanking is referred to by the name used by the Taipings, T'ien-ching 天京). The remaining short essays deal with Chu Chiu-t'ao 朱九濤, Huang Wan 黃畹, Chang Chia-hsiang 張嘉祥 and his alleged connection with Hung Hsiu-ch'üan, and instances of "following tradition and inheriting falsehoods" among other historical workers. (14,000)

2.5.19 Lo Erh-kang 羅爾綱, T'ai-p'ing t'ien-kuo shih-shih k'ao 太平天国史事考 (Authentication studies of historical facts on the Taiping Kingdom), Collected Essays on the History of the Taiping Kingdom, Vol. 2 (Peking: San-lien shu-tien, 1955), 359 pp.

The author's preface to this volume again refers to his method of textual research as that of the "Chia-ch'ing and Ch'ien-lung school" which, though it only sees the proverbial trees instead of the forest, is defended by Professor Lo as at least not presenting the trees as being something else. Marxism-Leninism is recommended for a view of the forest. Six of the seven articles in this volume deal with the following subjects: (1) the uprising at Chin-t'ien (which Professor Lo now dates Jan. 11, 1851); (2) the relationship between the Taiping Kingdom and the Heaven and Earth Society; (3) the controversy over the identity of Hung Ta-ch'üan; (4) the Taiping land system; (5) internal conflicts among the Taiping leaders; and (6) women in the Taiping kingdom. The seventh article, "A discussion with Professor Efimov of historical questions pertaining to the Taiping Kingdom" records the gist of a conversation between the author and Professor G. Efimov of the University of Leningrad in 1952, touching on these topics: (1) the revolutionary nature of the Taiping Kingdom; (2) a critique of its personalities; (3) the Taiping Kingdom and the Heaven and Earth Society; (4) the dating of the Taiping Kingdom. (14,000)

2.5.20 Lo Erh-kang 羅爾綱, T'ai-p'ing t'ien-kuo shih-liao pien-wei chi 太平天国史料辨偽集 (Collected writings on the detection of forgeries among historical materials on the Taiping Kingdom), Collected Essays on the History of the Taiping Kingdom, Vol. 3 (Peking: San-lien shu-tien, 1955), 138 pp., 2 plates.

The detection of forgeries among materials on the Taiping Kingdom has been a major concern of the author for over twenty years. In eight articles in this collection, he discusses several books and documents which he believes contain false accounts of the Taiping movement and/or are attributed to fictitious authors. Among the books are: (1) Chiang-nan ch'un-meng-an pi-chi 江南春夢庵筆記; (2) Hsün pi sui wen lu 盾鼻隨聞錄; and (3) T'ai-p'ing t'ien-kuo chan chi 太平天國戰紀. The forged documents include "A Proclamation of the Taiping Uprising" (T'ai-p'ing ch'i-i hsi-wen 太平起義檄文), and an "Imperial edict of the Taiping Kingdom" (T'ai-p'ing t'ien-kuo ch'ih-yü 太平天國敕諭). The authenticity of some 100 poems in an anthology called T'ai-p'ing t'ien-kuo shih wen ch'ao

太平天國詩文鈔 is disputed, as are some poems attributed to Shih Ta-k'ai 石達開. In addition, there is an article on some forged Taiping currency notes and coins. A small portion of the material in this collection appeared in an earlier volume called T'ai-p'ing t'ien-kuo shih pien-wei chi 太平天国史辨偽集 (Shanghai: Commercial Press, 1950), which treats not only spurious historical materials but also historical "facts" such as the alleged existence of Prince T'ien-te 天德王, Hung Ta-ch'üan 洪大全. (14,000)

2.5.21 Lo Erh-kang 羅爾綱, T'ien-li k'ao chi t'ien-li yü yin-yang li jih tui-chao piao 天曆考及天曆与陰陽曆日对照表 (A study of the Taiping calendar; with a table of corresponding dates according to the Taiping, lunar, and solar calendars), Collected Essays on the History of the Taiping Kingdom, Vol. 4 (Peking: San-lien shu-tien, 1955), 207 pp.

The first half of this volume discusses the principles of the Taiping calendar, its special features, the period when it was in use, the problems of determining corresponding dates in other calendars, and related problems on which Chinese historians have held differing opinions. The second half consists of a series of daily tables for the years 1852-68 which give the dates as reckoned by the Taiping, lunar, and solar calendars. (4,000)

2.5.22 Lo Erh-kang 羅爾綱, T'ai-p'ing t'ien-kuo shih-liao k'ao shih chi 太平天国史料考釋集 (Collected textual and annotative studies of historical materials on the Taiping Kingdom), Collected Essays on the History of the Taiping Kingdom, Vol. 5 (Peking: San-lien shu-tien, 1956), 344 pp., 5 plates.

This volume contains thirty-six brief articles which demonstrate Professor Lo's long and intensive textual study of Taiping source materials. Part I deals with Taiping books and manuscripts and includes, among other things, a bibliographical survey of the forty-two extant imprints of the Taiping Kingdom, and another article on five Taiping works which are known to have been produced but are not now extant. Part II concerns the "inscriptions and colophons" (*t'i-pa* 題跋) added to Taiping documents, the study of which is considered by Professor Lo to be fruitful in many instances. An important document in this group is a report by Lin Feng-hsiang 林鳳祥 and others on the Northern Expedition of 1853. Part III is devoted to studies of the works of Ch'ing officials and their followers. (6,000)

2.5.23 Lo Erh-kang 羅爾綱, *T'ai-p'ing t'ien-kuo wen-wu t'u shih* 太平天国文物圖釋 (Illustrations with explanatory texts of documents and objects of the Taiping Kingdom), Collected Essays on the Taiping Kingdom, Vol. 6 (Peking: San-lien shu-tien, 1956), 304 pp. (4,000)

2.5.24 Lo Erh-kang 羅爾綱, *T'ai-p'ing t'ien-kuo shih-chi tiao-ch'a chi* 太平天国史蹟調查集 (Investigations of historical relics of the Taiping Kingdom), Collected Essays on the History of the Taiping Kingdom, Vol. 7 (Peking: San-lien shu-tien, 1958), 393 pp. illus.

About half of this volume deals with the wall paintings discovered between 1951 and 1954 in Nanking, Shao-hsing, and other places which the author and other authorities have established as genuine relics of the Taiping Kingdom. The wall paintings in each locality are described in detail; unfortunately the plates which are included in this volume are very poor in quality.

This section also contains two articles on Taiping wall paintings which apparently have provoked some controversy among mainland specialists. The issues are: (1) were the wall paintings consciously promoted by the Taipings as a people's art form and are they "realistic" in presentation? (2) is the absence of human figures in these paintings dictated by religious belief? The remaining portion of the book concerns other relics of the Taiping era, e.g., stone tablets, manuscripts, and fragments of cannon supplied by Britain, France, or the United States to the Ch'ing government. The final article deals with the Taiping collection in the Nanking Museum. As an appendix, there is reproduced an article about Professor Lo's visit to Chin-t'ien, Kwangsi, in 1942. In a lengthy postface the author does some soul-searching with regard to the "new" versus the "old" textual study. (3,000)

2.5.25 Lo Erh-kang 羅爾綱, T'ai-p'ing t'ien-kuo ti li-hsiang-kuo--T'ien-ch'ao t'ien-mu chih-tu k'ao 太平天国的理想国--天朝田畝制度考 (The Taipings' utopia--a study of the Taiping land system; Shanghai: Commercial Press, 1950), 3 + 54 pp. (?)

2.5.26 Lo Erh-kang 羅爾綱, T'ai-p'ing t'ien-kuo hsin-chün ti yün-tung chan 太平天国新軍的運動战 (Guerrilla warfare of the Taiping Kingdom's New Army), rev. ed. (Shanghai: Commercial Press, 1955), 8 + 48 pp. (?)

2.5.27 Lo Erh-kang 羅爾綱, ed., T'ai-p'ing t'ien-kuo wen-hsuan 太平天国文選 (A literary collection on the Taiping Kingdom; Shanghai: Shang-hai jen-min ch'u-pan-she, 1956), 19 + 239 pp., 4 plates.

This collection consists of one or more examples of ten types of writing produced by Hung Hsiu-ch'üan and other leaders of the

Taiping Kingdom: essays, laws, edicts, proclamations, memorials, narrative, correspondence, epitaphs, autobiography, poems, and songs. Each item is accompanied by the editor's notes on the author, the source, the meaning, and historical context of the selection. Lo Erh-kang's interesting preface attempts, among other things, to justify the "religious cloak" of the Taiping Kingdom by referring to early Christianity as an "anti-Roman Empire and anti-Judaic clergy organization," and also by pointing to its primitive communal aspect. (5,500)

2.5.28 Hsieh Hsing-yao 謝興堯, T'ai-p'ing t'ien-kuo ch'ien-hou Kuang-hsi ti fan-Ch'ing yün-tung 太平天国前後廣西的反清運動 (The anti-Ch'ing movement in Kwangsi province before and after the Taiping Rebellion; Peking: San-lien shu-tien, 1950), 236 pp. (?)

2.5.29 Historical Research Seminar of North China University, ed., T'ai-p'ing t'ien-kuo ko-ming yun-tung lun-wen chi--Chin-t'ien ch'i-i pai-chou-nien chi-nien 太平天国革命運動論文集--金田起义百週年紀念 (Collected essays on the Taiping revolutionary movement commemorating the 100th anniversary of the Chin-t'ien uprising; Peking: San-lien shu-tien, 1950), 3 + 165 pp., 13 plates.

The commemorative article in this volume is contributed by Fan Wen-lan 范文瀾, who attributes the failure of the rebellion to strategic errors as well as to dissension and corruption among its leadership, and who praises the Taiping's "anti-feudal," "revolutionary" legacy to the Chinese Communists. The rest of the book contains articles on the organization of the Taiping Army, the commercial policy of the Taiping Kingdom, the "revolutionary people's" dealing with "foreign aggressors," Chinese and foreign "anti-revolutionary collusion" during the Taiping period, Hung Ta-ch'üan of the Heaven and Earth Society (in which Jung Meng-yuan 榮孟源 takes issue with Lo Erh-kang's thesis that Hung never existed), the Nien Army as a peasant movement in the North, and the Muslim rebellion led by Tu Wen-hsiu 杜文秀. The final article consists of short commentaries by Jung Meng-yuan on sixty-one Taiping publications. See the detailed review of this book by Mary Wright in Far Eastern Quarterly, 12.3:323-325 (May 1953). (15,000)

2.5.30 Kuang-hsi-sheng T'ai-p'ing-t'ien-kuo wen shih tiao-ch'a-t'uan 廣西省太平天国文史調查團 (Team of investigators of the culture and history of the Taiping uprising in Kwangsi province), T'ai-p'ing-t'ien-kuo ch'i-i tiao-ch'a pao-kao 太平天国起义調查报告 (A report of investigations on the Taiping uprising; Peking: San-lien shu-tien, 1956), 134 pp. + map and 14 plates.

Another example of the method of "historical research" increasingly used on the Chinese mainland. A team of ten-odd "historical, literary, and artistic workers," dispatched by the Kwangsi Provincial Department of Culture and History, covered large areas in Kwangsi near Chin-t'ien, the site of the first Taiping rising, and interviewed 291 persons, mostly old and poor peasants, whose knowledge of the Taiping Rebellion was admittedly little more than hearsay. On the basis of their recollections, and with only occasional reference to written documents, the investigating team has produced an account of Taiping activities in Kwangsi which is at best only an interesting concoction of local legends. About the only materials of historical value are the stone tablet inscriptions found in the villages, some texts of which are included in the appendix (pp. 99-112), along with the names, ages, occupations, and residences of the aged persons interviewed (pp. 113-134). (7,500)

2.5.31 Li Ch'un 酈純, T'ai-p'ing t'ien-kuo chih-tu ch'u-t'an 太平天国制度初探 (A preliminary inquiry into the institutions of the Taiping Kingdom; Peking: Jen-min ch'u-pan-she, 1956), 9 + 175 pp.

Aspects of the Taiping government studied in some detail are: (1) the abolition of private property, in theory and in practice; (2) the tax structure; (3) the public supply system, before and after the internal troubles of 1856; (4) the organization and functioning of the system of local officials; (5) the education and examination systems, and ideological changes; (6) social organization and communal living at the Taiping capital, Nanking. As in his second book on the Taiping Rebellion (2.5.33), the author relies heavily on the intelligence reports prepared for Tseng Kuo-fan (Tsei-ch'ing hui-tsuan 賊情彙纂) along with other records of the period. (5,000)

2.5.32 Chou T'un 周邨, T'ai-p'ing-chün tsai Yang-chou 太平軍在揚州 (The Taiping Army in Yangchow; Shanghai: Jen-min ch'u-pan-she, 1957), 4 + 61 pp., maps, plates.

The leading Chinese scholar on the Taiping Rebellion, Lo Erh-kang, has contributed a foreword to this book, from which we learn that the author collected materials on the Taipings while he was chief of the Propaganda Section of the Yangchow Local Committee of the Chinese Communist Party and directing a Taiping exhibition sponsored by the Yangchow Museum. Among his discoveries are two previously unknown contemporary works which are the main sources for his argument that the Taiping Army accomplished great things in Yangchow and won the support of the people. But, as Lo Erh-kang points out, since the Taiping Army held Yangchow for only nine months and was under constant seige, there is little concrete evidence that it was able to carry out the policies and measures it advocated. (15,000)

2.5.33 Li Ch'un 酈純, T'ai-p'ing t'ien-kuo kuan-chih chün-chih t'an-lueh 太平天国官制軍制探略 (A brief investigation into the bureaucratic system and military system of the Taiping Kingdom; Shanghai: Jen-min ch'u-pan-she, 1958), 4 + 136 pp.

A detailed and heavily documented outline of Taiping bureaucratic, military, and other organization, stressing the categories and prescribed functions of official personnel but without much analysis of how these institutions actually operated. The earlier phase of the Taiping Rebellion (until the internal upheaval of 1856) is given the fullest treatment (pp. 1-99), with proportionately less attention devoted to the later period. Among the principal sources employed is the Tsei-ch'ing hui-tsuan, an important intelligence report in the hands of Tseng Kuo-fan and the anti-Taiping forces. (6,500)

2.6 THE NIEN AND OTHER REBELLIONS

In addition to the Taiping movement, other mid-century rebellions have received considerable attention from Chinese Communist historians. These too are fitted into a revolutionary tradition of which the Chinese Communist victory is the inevitable outcome. This section includes materials on the Nien uprising, on the little-studied movement of Sung Ching-shih, and on activities of the secret societies contemporary with the Taiping and Nien uprisings. See also 2.7.2-3. For the mid-nineteenth century Muslim uprisings against the Manchu Dynasty, see sec. 1.3 above.

2.6.1 Chiang Ti 江地, Nien-chün shih ch'u t'an 捻軍史初探 (A preliminary exploration into the history of the Nien Army; Peking: San-lien shu-tien, 1956), 148 pp.

By way of an introduction, this volume proposes the following periodization for the history of the Nien uprising: (1) 1814-53: the Nien as a secret society; (2) 1853-55: the rise of the Nien Army, following the penetration of the Taiping Army into regions north of the Yangtze; (3) 1855-64: the height of the Nien rebellion, corresponding with the peak of Taiping activities; (4) 1864-66: the decline of the Nien Army, following the loss of its base in northern Anhwei, and paralleling the Taipings' loss of Nanking; (5) 1866-68: division into the Eastern and Western Nien Armies and the defeat of both. The text consists of five essays with these titles: (1) "The loss of the base in northern Anhwei and the martyrdom of Chang Lo-hsing 張洛行"; (2) "The meeting of the Northwest Taiping Army and the Nien Army"; (3) "On the battles of the Nien Army in the late period (i.e., after 1864)"; (4) "On the anti-Ch'ing struggles of the East Nien Army"; (5) "On the anti-Ch'ing struggles of the West Nien Army." Finally, there is a useful discussion of the source materials for Nien history, including some criticism of documents of the collection Nien-chün (2.6.2), and suggestions for investigating the contemporary Muslim uprisings in northwest China. The author notes that to date no materials emanating from the Nien themselves have come to light. (15,000)

2.6.2 Fan Wen-lan 范文瀾, Chien Po-tsan 翦伯贊, Nieh Ch'ung-ch'i 聶崇岐, Lin Shu-hui 林樹惠, and Wang Ch'i-chü 王其榘, eds., Nien Chün 捻軍 (The Nien Army), Modern Chinese Historical Materials Series, No. 3 (Shanghai: Shen-chou kuo-kuang she, 1953), 6 vols., 2,713 pp.

These materials on the Nien uprising are divided into three parts: overall accounts of the Nien by individual writers (Vol. 1); materials on the uprising extracted from local histories (Vols. 2-4); and letters, poems, essays, and the like, bearing on the Nien (Vols. 5 and 6). This rather inadequate classification and the absence of an index makes the locating of material on any specific topic a difficult matter. At the beginning of Vol. 1 there is a bibliography of 290 works, including 204 local histories, from which the editors have drawn their material (actually 302 works are quoted in the six volumes). No extracts have been included from the great Ch'ing official collection on the suppression of the rebellion, Ch'in-ting chiao-p'ing Nien-fei fang-lueh 欽定剿平捻匪方略 (Military plans for suppressing the Nien rebels, compiled under imperial auspices). See The Journal of Asian Studies, 17.1:76-80 (Nov. 1957), for a detailed review of this item by S. Y. Teng. (7,600)

2.6.3 Chiang Shih-jung 江世榮, ed., Nien-chün shih-liao ts'ung-k'an 捻軍史料叢刊 (A collection of historical materials on the Nien Army), No. 1 (Shanghai: Commercial Press, 1957), 9 + 152 pp.

This is the first in a series of three volumes on the Nien uprising. It contains three parts: military reports (dated in the eleventh and twelfth months of 1866), military intelligence (dated from the fourth to the eighth month of 1868), and selections from the diary of Neng Ching-chü 能靜居 (1858-89) between 1860 and 1868. Like the newly discovered letters of Li Hung-chang (which constitute No. 2 of this series), the materials for this volume did not come to light until 1954 when the Kiangsu Provincial Commission on the Management of Cultural Materials took over the holdings of the T'ai-ts'ang-hsien 太倉县 library. The documents are accompanied by ample explanatory notes by the editor. This collection and the two which follow supplement the materials in 2.6.2. (11,000)

2.6.4 Chiang Shih-jung 江世榮, Nien-chun shih-liao ts'ung-k'an 捻軍史料叢刊 (A collection of historical materials on the Nien Army), No. 2 (Shanghai: Commercial Press, 1957), 31 + 164 pp.

Two hundred and fourteen letters, hitherto unknown, written by Li Hung-chang during the Nien Rebellion between 1866 and 1868. The table of contents lists the full total of 312 letters which are discovered in 1954 (see the report by Chiang Shih-jung in Li-shih yen-chiu [Historical studies], p. 52 [March 1955]); and persons and places referred to are identified in a total of 690 notes. The third volume in this series was not seen. (4,500)

2.6.5 Nieh Ch'ung-ch'i 聶崇岐, ed., Nien-chün tzu-liao pieh-chi 捻軍資料別集 (A further collection of materials on the Nien Army; Shanghai: Jen-min ch'u-pan-she, 1958), 14 + 348 pp.

Having participated as editor and compiler of Nien-chün (2.6.2), Mr. Nieh has collected further materials on this subject from public and private documents compiled after 1821. Thirty-two compilations are listed at the beginning of this book, together with a summary of each author's official career. Selections from these compilations include memorials, proclamations, essays, letters, biographical sketches, and diaries, and deal with the activities of the Nien forces roughly over the years 1800-70. The editor's preface sees the Nien uprising as a direct descendant of the White Lotus rebellion of the late eighteenth century. (3,000)

2.6.6 Office of Ming and Ch'ing Archives, Bureau of National Archives, comp., Sung Ching-shih tang-an shih-liao 宋景詩檔案史料

(Archival materials relating to Sung Ching-shih; Peking: Chung-hua shu-chü, 1959), 20 + 371 pp.

A collection of 287 previously unpublished documents, largely from the archives of the Grand Council, dealing with the little-studied peasant rebellion led by Sung Ching-shih which erupted in Shantung in 1861. Sung was a leader of the "Black Standard" (Hei-ch'i 黑旗), one of the five divisions of the White Lotus Society. This publication, and items 2.6.7-8, illustrate the dominant Chinese Communist interest in peasant risings, although the materials actually collected here are records left by the "feudal rulers"—memorials by officials, imperial edicts, and the like, dealing with Sung Ching-shih's activities as well as with the White Lotus Society in general. The period covered by the bulk of these documents is 1861-63; an appendix includes one item dated 1867, and three dated 1871. (4,200)

2.6.7 Ch'en Pai-ch'en 陳白塵 et al., Sung Ching-shih li-shih tiao-ch'a chi 宋景詩歷史調查記 (An account of the investigations into the history of Sung Ching-shih; Peking: Jen-min ch'u-pan-she, 1957), 7 + 280 pp., 8 plates.

This volume is a product of a two-month (March to May 1952) "field study" by a team of eleven persons in 163 villages in eight hsien of Shanghai province, the birthplace of Sung Ching-shih (1824-?). A total of 724 persons, mostly aged peasants in their seventies and eighties, were interviewed in an effort to obtain information on the life and times of this peasant rebel. (The name, age, and occupation of all the interviewees are given in an appendix.) The result is a narrative, in seventeen chapters, on Sung as a peasant and as the military leader of a peasant uprising in Shantung contemporary with the Taiping Rebellion. Mr. Ch'en, who alone wrote up the findings of the eleven-man team, appears to have consulted many historical documents and a number of scholarly works as indicated by his bibliography (pp. 279-280), but he claims that the value of this volume lies primarily in its "predominant content of peasant folklore." It is, however, difficult to see that the interviews contributed much more than unverifiable local legends about some of the participants in Sung's rebellion. (15,000)

2.6.8 Cheng T'ien-t'ing 鄭天挺 et al., eds., Sung Ching-shih ch'i-i shih-liao 宋景詩起义史料 (Historical materials on the uprising of Sung Ching-shih; Peking: Chung-hua shu-chü, 1954), 2 + 176 pp.

This is a revised edition of Sung Ching-shih shih-liao 宋景詩史料, published by K'ai-ming shu-tien, Peking, in 1953. The principal materials are taken from three previously unpublished manuscripts, one of which consists of the memorials of Ch'ung-hou 崇厚, who from 1861 to 1870 was superintendent of trade for the northern ports of China. Other selections are from about a dozen official Ch'ing records, e.g., Ch'ing shih-lu and Shan-tung chün-hsing chi-lueh 山東軍興紀略. (The extracts from hsien gazetteers included in the earlier edition are now omitted.) These selections are arranged in six chapters, in chronological order and by areas in which Sung's peasant rebellion was active. A short introductory account of Sung Ching-shih contrasts him with Wu Hsün 武訓 (also a native of Shantung and contemporary with Sung), who, unlike him, "surrendered" to the ruling landlord class. A list of personal names is appended (pp. 171-176) which identifies individuals in both the peasant and the Ch'ing forces. (7,000)

2.6.9 Tsinan Branch of the Chinese Historical Association, ed., Shan-tung chin-tai shih tzu-liao 山東近代史資料 (Source materials on the modern history of Shantung; Tsinan: Shan-tung jen-min ch'u-pan-she, 1957), Vol. 1, 6 + 280 pp., 8 plates.

The contents of this volume include local materials in manuscript or printed form, selections from gazetteers, official documents, and pi-chi, plus a record of visits to historical sites and interviews with local residents made by the Editorial Committee on Shantung Modern History of the Shantung Historical Association. These materials deal with rebellions in Shantung in the 1850's and 1860's, most of which were related in some way to the Taipings, the Nien, or the White Lotus sect. The editors include some introductory and explanatory notes, but they do not in general give enough guidance to local figures and events to make their materials fully useful. (4,100)

2.6.10 Preparatory Committee of the Shanghai Institute of Historical Research, Chinese Academy of Sciences, comp., Shang-hai Hsiao-tao-hui ch'i-i shih-liao hui-pien 上海小刀会起义史料汇編 (Historical materials on the uprising of the "Small-knife Society" in Shanghai; Shanghai: Shang-hai jen-min ch'u-pan-she, 1958), 40 + 1032 pp., 13 plates.

Following a general account of the activities of the "Small-knife Society" and a chronological table of major events (Feb. 20, 1840-Feb. 17, 1855) prepared by the editors, the text of this collection includes: (1) twenty-five documents of the society, mainly proclamations, about half of which are obtained from Chinese sources, while the remainder are translated into Chinese from the North China Herald (NCH) for the years 1853-54; (2) accounts of the Small-knife Society's uprising and reports on the fighting in Shanghai (the bulk of this material is taken from the NCH); (3) documents on the Ch'ing government's suppression of the rebellion in the form of intelligence reports, memorials, and the like, and translations from the NCH of some Ch'ing official proclamations and reports; (4) foreign consular notices, letters, newspaper articles, and the like, which indicate the position of the "foreign aggressors" with regard to the Small-knife Society (including translations from the NCH, and substantial selections from other English-language sources); (5) miscellaneous selections (e.g., gazetteers, poems, memoirs) dealing with the Small-knife Society and with other uprisings in the vicinity of Shanghai in this period. These materials are carefully edited, the sources are clearly indicated, and the editors have supplied explanatory notes for the text as well as general interpretive remarks. (2,500)

2.6.11 Nieh Ch'ung-ch'i 聶崇岐, ed., Chin-ch'ien hui tzu-liao 金錢会資料 (Material on the "Golden Coin Society"; Shanghai: Shanghai jen-min ch'u-pan-she, 8 + 208 pp.

This relatively little-known anti-Manchu secret society flourished briefly toward the end of the Taiping Rebellion in southeastern Chekiang and northeastern Fukien. Accounts of its activities and their suppression are selected from the writings of scholars and officials in that area, including Tso Tsung-t'ang, and also from several local gazetteers. (3,000)

2.7 RESTORATION, SELF-STRENGTHENING, AND THE REFORM MOVEMENT OF 1898

For the Chinese Communist historians, as for the rest of us, the the three decades after the suppression of the Taiping Rebellion remain *terra incognita*. Compared with the plethora of compilations and accounts on the Taiping and other mid-nineteenth century rebellions, relatively little work has been published on the T'ung-chih Restoration, on the "self-strengthening" efforts of the 1860's and 1870's, or on the general process of Westernization and the spread of Western influence in these last decades of the century. The general framework within which this period is viewed by the mainland historians is one which stresses the treacherous combination of the feudal Manchu ruling class with the aggressive foreign imperialists so that, on the one hand, Ch'ing rule is maintained while, on the other, the imperialists are duly paid for their support of the reactionary regime. In the current version of nineteenth-century history there are no greater "enemies" of the Chinese people than Tseng Kuo-fan and Li Hung-chang. These two men, who led the suppression of the Taipings, were among the first to acknowledge China's need to adopt Western arms and military and scientific technology if she were to survive in the new world to which she was introduced by the Opium War and its aftermath. Such attention as has been given to this period has tended to concentrate on the industrialization efforts undertaken by Li Hung-chang and others (see 2.7.1, and sec. 4.4) or on the publication of source materials. The newly published Tseng Kuo-fan letters (2.7.2) and a new edition of Liu K'un-i's collected works (2.7.3), a rather scarce item, are especially valuable sources for the study of this period.

A good deal more has been done with the Reform Movement of 1898, which forms an essential link in the Chinese Communist construction of an historical pedigree for their own seizure of power. Despite the depressive effect which the "monopolistic" enterprises undertaken by the self-strengthening leaders had on the growth of private capitalistic industry and a native bourgeoisie, nothing could halt the inexorable progress of economic forces. Some truly capitalist enterprise was able to grow despite imperialist aggression and feudal oppression, and this, the mainland historians assert, is the "material basis" for the "bourgeois" program of the reformers of 1898. The failure of the "bourgeoisie" in 1898, in 1911, and after, of course is taken as evidence that only a popular political movement led by the proletariat (i.e. the Communist Party) could bring fundamental change into Chinese society. Items 2.7.5-8 are general surveys and studies of 1898. Liang Ch'i-ch'ao's own account of the events of that year is of course a first-hand source, rather than a scholarly study; it is followed by two extremely important collections of source materials (2.7.10 and 2.7.11). It is of interest that among the leaders of the reform movement Liang Ch'i-ch'ao has received relatively little attention from the mainland historians, perhaps because of his later political and scholarly career. Items 2.7.12-19 are good examples of recent studies of K'ang Yu-wei and T'an Ssu-t'ung and of the new editions of their works. See also 2.2.9, and 5.2.1-5.

2.7.1 Mou An-shih 牟安世, Yang-wu yun-tung 洋務運動 (The "foreign matters movement"; Shanghai: Shang-hai jen-min ch'u-pan-she, 1956), 4 + 230 pp.

On the surface, this is a detailed and heavily documented study of the so-called "foreign matters movement," i.e., the industrialization effort undertaken by leading provincial officials in the last half of the nineteenth century; but it is often quite thin in content when describing specific enterprises. This account divides the "movement" into three stages: (1) the establishment of arsenals and an arms industry, 1860-72; (2) the establishment of non-military enterprises in support of these military industries, 1872-85; and (3) the establishment of the Peiyang Navy and of iron foundries (i.e., heavier industry), 1885-94. The introductory discussion of China's mid-nineteenth-century domestic condition and foreign relations strongly asserts the essentially "reactionary" and "treasonous" character of the movement. The outcome and impact are seen as impeding the growth of Chinese native industry and shaping China into a "semi-feudal and semi-colonial" society. Appended materials include a chronological table of major events from 1860 to 1894, a list of foreign names opposite their Chinese transliterations, and a brief bibliography which includes Marxist, Ch'ing, and English works. (23,000)

2.7.2 Chiang Shih-jung 江世榮, ed., Tseng Kuo-fan wei k'an hsin-kao 曾國藩未刊信稿 (Unpublished letters of Tseng Kuo-fan; Shanghai: Chung-hua shu-chü, 1959), 25 + 393 pp.

This volume consists of 461 hitherto unpublished letters of Tseng Kuo-fan, written between 1861 and 1871, all having some bearing on major events of that period. (In recent years a large collection of Tseng's letters was discovered, 1006 of which had never been published. Besides the present collection, 192 of these letters are included in 2.6.2.) Appended are ninety-four letters from other Ch'ing officials to Tseng (pp. 302-373), plus selections from Neng Ching-chü jih-chi 能靜居日記, the diary of Chao Lieh-wen 趙烈文, a protegé of Tseng (pp. 374-393). (3,400)

2.7.3 Third Office of the Institute of Historical Research, Chinese Academy of Sciences, ed., Liu K'un-i i chi 劉坤一遺集 (A posthumous collection of Liu K'un-i's writings; Peking: Chung-hua shu-chü, 1959), 6 vols., 100 + 2813 pp.

The materials in this collection are reprinted from Liu Chung-ch'eng-kung i chi 劉忠誠公遺集 (1921, 67 chüan). With the exception of five chüan of literary writings, all the papers left by Liu (1829-1902), who long was a key official in the Yangtze provinces, are included and arranged under the usual categories: memorials, telegraphed memorials, letters, telegrams, and official correspondence. These papers, as the editors summarize them, concern: (1) "people's revolutions" and anti-Ch'ing struggles, especially the Taiping Army and the Heaven and Earth Society; (2) negotiations with foreign countries; (3) military training and coastal defense; (4) the Sino-Japanese war; (5) the Boxers, the pacification of South China, and the "Sino-Russian secret treaty"; and (6) finance (e.g., grain tribute, problems of coinage, ordnance and military supplies). The editors have contributed a brief commentary on Liu's official career, a table of contents for each of the volumes, punctuation and paragraphing and indications of obvious or possible errors and omissions. (1,000)

2.7.4 Hu Pin 胡濱, Mai-kuo-tsei Li Hung-chang 賣国賊李鴻章 (Li Hung-chang, the traitor; Shanghai: Hsin chih-shih ch'u-pan-she, 1955), 98 pp.

This study of the official career of Li Hung-chang, one of the arch villains of current mainland historiography, deals with his suppression of the Taiping and Nien rebellions, his policy of "self-strengthening," his "sell-out" negotiations with Japan, Britain, Russia, and France through the period of the Sino-Japanese War, and other "treasonous" acts or "plots" up to the end of the Boxer rebellion and Li's death in 1901. There is little new material on Li, but the extreme nationalistic tone in dealing with nineteenth-century history is of interest. (28,000)

2.7.5 Hu Pin 胡濱, Wu-hsü pien-fa 戊戌変法 (The Reform of 1898); Shanghai: Hsin chih-shih ch'u-pan-she, 1956), 112 pp.

This small, semi-popular volume includes convenient brief accounts of the life and thought of four major "bourgeois reformers" (K'ang Yu-wei, Liang Ch'i-ch'ao, T'an Ssu-t'ung, and Yen Fu 嚴復) and of Chang Chih-tung 張之洞, who is singled out among the "yang-wu" school as an opponent of the reformers. These accounts are sandwiched between a description of imperialist aggression in nineteenth-century China and the development of incipient capitalism on the one hand, and a narrative of the reform movement and its collapse, on the other. The volume adds little to our knowledge of the events of 1898, and is largely an exegesis of Mao Tse-tung's dictum that the reform movement was inspired by one part of the capitalist class and the more enlightened landlords, and that it failed because it was separated from any political movement of the masses. (22,000)

2.7.6 T'ang Chih-chün 湯志鈞, Wu-hsü pien-fa shih lun ts'ung 戊戌變法史論叢 (Collected essays on the history of the 1898 Reform; Wuhan: Hu-pei jen-min ch'u-pan-she, 1957), 322 pp.

As the titles of the fourteen essays will show, this collection provides an extensive treatment of the reform and related subjects: (1) "An analysis of the factions within the Ch'ing ruling class at the time of the Reform"; (2) "The Reform and American imperialism"; (3) "'Progressive' thought before the Reform"; (4) "The 'modern text' school of Ch'ang-chou in the Ch'ing period and the Reform"; (5) "The 'yang-wu movement' and the Reform"; (6) "Weng T'ung-ho 翁同龢 and the Reform"; (7) "The origins of K'ang Yu-wei's reformist thought"; (8) "K'ang Yu-wei's ta-t'ung ideology and the Ta-t'ung shu 大同書"; (9) "A study of the chronology of K'ang Yu-wei's Li yun chu 禮運注"; (10) "K'ang Yu-wei's reform proposals and Kuang-hsü's reform edicts"; (11) "Scholarly associations, newspapers, and periodicals during the Reform"; (12) "Concerning the 'secret edicts' of Kuang-hsü"; (13) "The six martyrs of 1898"; and (14) "T'ang Ts'ai-ch'ang 唐才常 and the 'Independent Army' uprising." These articles are scholarly, well-documented, and on the whole have an analytical approach. Communist dogma is rather sparingly used, except in the article on American imperialism. The tenth essay includes a listing of the reform proposals and imperial edicts in tabular form, and can be used as a ready reference. (4,400)

2.7.7 Hou Wai-lu 侯外廬, chief ed., Wu-hsü pien-fa liu-shih chou-nien chi-nien chi 戊戌變法六十週年紀念集 (A volume commemorating the sixtieth anniversary of the 1898 Reform; Peking: K'o-hsueh ch'u-pan-she, 1958), 82 pp., illus. (?)

2.7.8 Wu Yü-chang 吳玉章 et al., Wu-hsü pien-fa liu-shih chou-nien chi-nien lun-wen chi 戊戌變法六十週年紀念論文集 (Collected essays on the sixtieth anniversary of the Reform of 1898; Peking: Chung-hua shu-chü, 1958), 1 + 101 pp. (4,400)

2.7.9 Liang Ch'i-ch'ao 梁啟超, Wu-hsü cheng-pien chi 戊戌政變記 (A record of the 1898 coup d'état; Peking: Chung-hua shu-chü, 1954), 2 + 157 pp.

A reprint of Liang's account of 1898 from the collection Yin-ping-shih ho-chi 飲冰室合集, interestingly enough without the explanatory prefatory remarks that are common for reprints of older works published in the Chinese People's Republic. (15,000)

2.7.10 Chien Po-tsan 翦伯贊 et al., eds., Wu-hsü pien-fa 戊戌變法 (The Reform of 1898), Modern Chinese Historical Materials Series, No. 8 (Shanghai: Shen-chou kuo-kuang ch'u-pan-she, 1953), 4 vols., 2,491 pp.

Materials from 175 works, included either in entirety or in part, relating to the abortive reform movement of 1898. Of particular interest are nine important hitherto unpublished manuscripts, and extracts from hard-to-find contemporary newspapers and magazines. Much of the other material is relatively well-known and includes edicts, memorials, and letters by the leading reformers and their opponents. Vol. 4 contains biographical materials, a collection of original sources on study groups and reform associations, a chronology of the Hundred Days, and a descriptive bibliography of 296 items, which in general neglects Japanese and Western writings. See the detailed review of this collection by Chaoying Fang in The Journal of Asian Studies, 17.1:99-105 (Nov. 1957). (9,500)

2.7.11 Office of Ming and Ch'ing Archives, Bureau of National Archives, comp., Wu-hsü pien-fa tang-an shih-liao 戊戌變法檔案史料 (Archival materials on the Reform of 1898; Peking: Chung-hua shu-chü, 1958), 21 + 524 pp.

This volume supplements the memorials from the Ch'ing archives concerning the 1898 reform movement printed in the 1953 collection, Wu-hsü pien-fa (2.7.10). It contains 303 proposals for reform arranged by topic: general proposals, recommendations of new personnel, changes in the bureaucratic structure, changes in the examination system, schools and study abroad, new army and militia, agricultural and commercial affairs, banking and currency, mining and railroad construction, translation bureaus, and "others." Three appendices (pp. 509-524) list: (1) documents included in 2.7.10 and omitted from this collection; (2) titles and dates of 226 memorials available in the archives but not included in either this volume or the earlier collection; and (3) a memorial from K'ang Yu-wei to P'u-i 溥儀 (the Hsuan-t'ung Emperor) dated 1927. (2,100)

2.7.12 Li Tse-hou 李泽厚, K'ang Yu-wei T'an Ssu-t'ung ssu-hsiang yen-chiu 康有為譚嗣同思想研究 (A study of the thought of K'ang Yu-wei and T'an Ssu-t'ung; Shanghai: Shang-hai jen-min ch'u-pan-she, 1958), 235 pp. (3,000)

2.7.13 Sung Yün-pin 宋雲彬, K'ang Yu-wei 康有為 (Peking: San-lien shu-tien, 1955, first published 1951), 7 + 131 pp. (28,000)

2.7.14 K'ang Yu-wei 康有為, Ta-t'ung shu 大同書 (Peking: Ku-chi ch'u-pan-she, 1956), 6 + 301 pp.

A reprint of one of K'ang's major works, this volume has been carefully edited from the first complete published edition (1935), with the aid of a manuscript copy in the possession of the K'ang family. The publisher's note gives K'ang only mild praise ("his progressive aspect cannot be denied"), while it denounces the "reformism" of which this work is a "highest manifestation." (3,000)

2.7.15 Yang Cheng-tien 楊正典, T'an Ssu-t'ung -- chin-tai Chung-kuo ch'i-meng ssu-hsiang-chia 譚嗣同--近代中国啟蒙思想家 (T'an Ssu-t'ung -- an instructive thinker of contemporary China; Wuhan: Hu-pei jen-min ch'u-pan-she, 1955), 60 pp., 1 plate. (6,000)

2.7.16 Yang T'ing-fu 楊廷福, T'an Ssu-t'ung nien-p'u 譚嗣同年譜 (A chronology of the life of T'an Ssu-t'ung; Peking: San-lien shu-tien, 1957), 127 pp. (5,300)

2.7.17 T'an Ssu-t'ung ch'üan-chi 譚嗣同全集 (A complete collection of T'an Ssu-t'ung's works; Peking: San-lien shu-tien, 1954), 16 + 534 pp., 4 plates.

This is the most complete published collection of writings by T'an Ssu-t'ung (1866-98), gathered from both public and private sources and carefully edited. Part I, treatises, contains T'an's important work, Jen hsueh 仁學, and twenty-odd short discourses on various subjects. Part II, narrative essays, contains some prefaces, biographies, as well as T'an's autobiography at thirty, "San-shih tzu-chi" 三十自紀. Part III, letters, includes his "proposal for the promotion of mathematics" and some hitherto unpublished letters (to Liang Ch'i-ch'ao and others). Part IV consists of poems. An appendix contains a biography of T'an by Liang Ch'i-ch'ao, a dirge by K'ang Yu-wei, and two prefaces, one to Jen hsueh by Liang and one by Ou-yang Yü-ch'ien 歐陽予倩 (grandson of T'an's teacher Ou-yang Pien-chiang 歐陽辦彊). The sources for this compilation are given in a postscript by the editor, Ts'ai Shang-ssu 蔡尚思, which is dated October 1, 1948. (10,000)

2.7.18 Wen Ts'ao 文操, ed., T'an Ssu-t'ung chen chi 譚嗣同真蹟 (Authentic autographs of T'an Ssu-t'ung; Shanghai: Shang-hai ch'u-pan kung-ssu, 1955), 2 + 166 pp. (folio). (650)

2.7.19 T'an Ssu-t'ung 譚嗣同, Jen hsueh 仁學 (A study of benevolence; Peking: Chung-hua shu-chü, 1958), 2 + 83 pp., 1 plate.

Reprint of a 1911 edition published by the Kuo-min pao-she ch'u-yang hsueh-sheng pien-chi-so 国民报社出洋學生編輯所, with punctuation marks, adopted from the version included in T'an Ssu-t'ung ch'üan-chi (2.7.17). The publisher's note makes the usual remarks about T'an's dual manifestation of hatred for the feudal system and lack of faith in the masses. (3,000)

2.8 THE BOXER REBELLION

The Boxer Rebellion, like the Taiping Rebellion, is now a hallowed part of the revolutionary tradition leading inexorably to the establishment of the People's Republic of China in 1949. Even to use the term "rebellion" with reference to either of the two movements is to display an "imperialist's" fear and hatred of these popular uprisings. The Boxer movement, say the mainland historians, was both an uprising of the exploited peasantry against the feudal ruling class and a popular anti-imperialist movement directed against the foreign powers who had begun to "carve up" China after the Sino-Japanese War of 1894-95. The Manchu Dynasty and the Empress Dowager were seriously threatened from two directions. Tz'u Hsi's declaration of war against the Western powers was, of course, a cunning policy aimed at getting rid of the Boxers by embroiling them with superior foreign troops. But, the Communist historians hold, this was only a secondary motive. Mao Tse-tung has stated that notwithstanding the importance of the internal class struggle, the "principal contradiction" in modern Chinese history is that between imperialism and the Chinese nation. The Empress Dowager's decision for war can, therefore, be interpreted as a positive response of the court to the imperialist threat to the existence of the dynasty, a decision that was made under the impact of a popular anti-imperialist movement. Again, the Boxers failed because they lacked proletarian leadership and organization, and because of savage repression by the imperialist armies who occupied North China. The strength of the anti-foreign feeling revealed in the Boxer movement, however, made it evident to the imperialist powers that they could not proceed further with dividing China among themselves. Thus, the Boxer uprising is currently interpreted as having saved China from becoming another Africa or an India, a colony of the Western powers. The first three items below are typical secondary studies of the Boxer uprising. They are followed by four valuable collections of sources, including a considerable amount of previously unpublished material. See also 2.2.9, 2.7.3 and 2.9.14.

2.8.1 Ming Ch'ing 銘青 I-ho t'uan 义和团 (The Boxers; Shanghai: Shih-tai shu-chü, 1950), 2 + 96 pp. (?)

2.8.2 Shih-hsueh shuang-chou-k'an she 史學双周刊社 ("Historical Biweekly" society), ed., I-ho-t'uan yun-tung shih lun ts'ung 义和团运动史論丛 (Collected essays on the history of the Boxer movement; Peking: San-lien shu-tien, 1956), 135 pp.

Fifteen short articles by ten writers (several of whom are members of the Institute of Historical Research) carried in mainland newspapers, between 1951 and 1955. Their titles are: (1) "Yuan Shih-k'ai's suppression of the Boxers in Shantung"; (2) "The so-called "Yangtze Compact" (Tung-nan hu-pao 東南互保)"; (3) "The Boxer Protocol and imperialism"; (4) "The Boxer Protocol and the so-called legation quarters"; (5) "The Boxer Indemnity--a picture of imperialists fighting over their loot"; (6) "The Boxer Indemnity and peasants' resistance to payments in 1902"; (7) "On the continued struggles of the Boxers in central Hopei after the Boxer Protocol"; (8) "The Boxers' posters"; (9) "Background of U.S. imperialism's 'Open Door policy'"; (10) "The content and aggressive nature of U.S. imperialism's 'Open Door policy'"; (11) "The American plot to seize San-sha-wan 三沙灣 (Fukien)"; (12) "Diary of an Indian soldier during the Boxer movement"; (13) "Criminal acts of the U.S. missionary W.A.P. Martin during the Boxer movement"; (14) "Plundering by the French Catholic priest Favier in 1900-1901"; (15) "Two Catholic spies--Favier and Anzer." As the titles indicate, the theme of imperialist aggression is pursued from many angles. Western sources are liberally cited, if only to be refuted (e.g., H.M. Vinacke's and Tyler Dennett's interpretations of the Open Door). (10,000)

2.8.3 Chin Chia-jui 金家瑞, I-ho-t'uan yun-tung 义和团运动 (The Boxer movement; Shanghai: Shang-hai jen-min ch'u-pan-she, 1957), 2 + 187 pp., 1 map.

A detailed account of the Boxer Rebellion of 1900 from its outbreak up to the conclusion of the Boxer Protocol. The text is amply documented with Chinese and Western sources, including public and private accounts, and also cites several Japanese works. The treatment is fully orthodox; in fact, the author states that some of his phraseology is borrowed from Hu Sheng's Ti-kuo chu-i yü Chung-kuo cheng-chih (1.4.5). An appendix (pp. 182-185) deals with "Some questions concerning the 'Boxer Protocol' indemnities," e.g., the voluntary forfeit of indemnity funds by the U.S.S.R. (non-exploitation) and the U.S. Boxer Indemnity Fund ("cultural aggression"). The author was one of the editors of the 4-vol. documentary collection I-ho-t'uan (2.8.4). (14,000)

2.8.4 Chien Po-tsan 翦伯贊 et al., eds., I-ho-t'uan 义和团 (The Boxers), Modern Chinese Historical Materials Series, No. 9 (Shanghai: Shen-chou kuo-kuang she, 1953, originally published March 1951), 4 vols., 2,243 pp.

Most of the fifty-six items in this collection have previously appeared in print. There are, however, five brief previously unpublished manuscripts and seven items compiled by the editors themselves. These last include documents relating to the preservation of peace in South China, over four hundred edicts copied from the Ch'ing shih-lu 清實錄, contemporary editorials and articles from Chinese and foreign newspapers, and a valuable annotated bibliography (Vol. 4, pp. 527-623). See the detailed reviews of this collection by Chaoying Fang in the Far Eastern Quarterly, 12.3:327-329 (May 1953); and by Lienche Fang in The

Journal of Asian Studies, 17.1:105-109 (Nov. 1957). (12,000)

2.8.5 Office of Ming and Ch'ing Archives, National Archives, ed., I-ho-t'uan tang-an shih-liao 义和团檔案史料 (Archival materials on the Boxer Rebellion; Peking: Chung-hua shu-chü, 1959), 2 vols., 100 + 1346 pp., 4 plates.

These volumes mark the first appearance of a complete collection of archival material on the Boxer Rebellion, including memorials with imperial rescripts originally kept in the Ch'ing palaces and various documents received or despatched by the Grand Council. A number of edicts omitted from the Ch'ing shih-lu are included, and some altered shih-lu documents given in their original texts. Among the palace materials are also reports from various yamen and from emissaries in foreign countries bearing on the Boxer crisis. These documents, totaling 800,000 characters, begin on July 2, 1896 and end on January 18, 1902. Repetitions of the texts of edicts within memorials have been deleted by the editors, and notes have been added directing the reader to the appropriate edict. (2,300)

2.8.6 Third Office, Institute of Historical Research, Chinese Academy of Sciences, ed., Keng-tzu chi-shih 庚子記事 (Chronicles of the Boxer Rebellion; Peking: K'o-hsueh ch'u-pan-she, 1959), 266 pp.

This volume consists of five chronicles or diaries dealing with the Boxer affair in 1900, mainly in Peking. One of these has been published before, but it is rarely seen: "Kao Nan jih-chi" 高枏日記 (The diary of Kao Nan). The other four are published here for the first time from manuscripts in the possession of the Institute of Historical Research. Nothing is known about the authors except that they were apparently official-gentry who lived in Peking at the time of the Boxer disturbances, and mingled their eyewitness accounts with hearsay. (The one exception is Hua Hsuan-lan 華學瀾, a chü-jen of 1885 and a minor official in the capital.) The editors have omitted from the present volume those materials included in the five chronicles (e.g., public notices) which can be found in the major recent collection on the Boxers (2.8.4). They have added useful editorial notes as well as an introduction which describes the Boxer Rebellion as "the people's anti-imperialist struggle, without modern proletarian leadership." It dismisses a reference in "Kao T'ung's diary" to "the people in the street asking American officers to remain" by trying to show that the U.S. was a major perpetrator of crimes during the Boxer affair. (3,200)

2.8.7 A Ying 阿英, ed., Keng-tzu shih-pien wen-hsueh chi 庚子事變文學集 (A literary collection on the Boxer incident; Peking: Chung-hua shu-chü, 1959), 2 vols., 43 + 1152 pp., 8 plates.

The editor notes in his introduction that the literary materials on the Boxer Rebellion are more numerous than for previous anti-foreign movements. This collection is therefore more substantial than items 2.4.7, 2.9.4, and 2.9.11. Aside from the usual categories of verse and novel, there are selections of "ballads" (shuo-ch'ang 說唱, lit., "recitation and singing"), some of which were written for stage production. A number of newspaper articles are also included among the essays. While the editor's dominant concern is with themes of patriotism and anti-foreign sentiments, the viewpoints represented by the selections are more diverse (e.g., Lin Shu 林紓 is decidedly anti-Boxer, and Pierre Loti--in translation, the only non-Chinese author included--is, one might say, uninvolved). The editor's introduction, in addition to making the Communist ideological point with regard to class identification, contains some useful information on various literary products of the Boxer period. (2,500)

2.9 LATE CH'ING FOREIGN RELATIONS

The emphasis in the Chinese Communist treatment of late Ch'ing foreign relations is, of course, on "imperialist aggression" against China. To be sure, the Western record in China in these years is often a regrettable one, but the crude analysis provided by the Chinese Communist historians seems just as misleading as the sometimes superficial Western language treatments of this period which have concentrated on the formal diplomatic record. A satisfactory scholarly treatment of China in the age of imperialism which takes account both of China's domestic history and of the international situation, is lacking in any language. At the same time, it is one of the most critical desiderata in the field of Chinese history. If China and the West are ever to come together in the present, there must also be some reconciliation of widely divergent views of what happened in the past.

Item 2.9.1 is a good example of current treatment of China's relations with the powers in the late nineteenth century; see also the general surveys in sec. 1.4 above. The remainder of this section includes works on the Sino-French War of 1884-85, the Sino-Japanese War of 1894-95, and four titles (2.9.15-18) that deal with American "aggression" against China. Note particularly the important collections of source materials on the Sino-French War and the Sino-Japanese War (2.9.2, 2.9.10), and the valuable series of documents from the archives of the Imperial Maritime Customs (2.9.3, 2.9.5-6, 2.9.12-13). Item 2.9.14 is a useful addition to published sources on late Ch'ing history, including foreign relations. See also 2.2.9 and 2.7.3.

2.9.1 Hu Pin 胡濱, Shih-chiu shih-chi mo-yeh ti-kuo-chu-i cheng-to Chung-kuo ch'üan-i shih 十九世紀末葉帝国主义争奪中国

权益史 (A history of imperialist strife over the plunder of Chinese rights and interests at the end of the nineteenth century; Peking: San-lien shu-tien, 1957), 2 + 220 pp.

Although it uses sources such as Weng T'ung-ho's 翁同龢 diary, Sheng Hsuan-huai's 盛宣懷 papers, the Ch'ing-chi wai-chiao shih-liao 清季外交史料, in addition to the standard Western-language sources, this volume adds little new factual material to existing accounts of its subject. The potential of the Chinese sources is barely touched; and note that no use is made of Die Grosse Politik, an indispensable source if one is attempting to demonstrate, as is the author, that England and the United States were no less desirous of making political and economic gains in China than were Russia, France, and Germany. That the Western record in China in the years 1894-99 is sometimes an unenviable one is indubitable; but the crude explanations here offered do not help to explain either imperialism or the Chinese response to it. A bibliography is appended (pp. 214-217). (6,000)

2.9.2 Shao Hsün-cheng 邵循正 et al., eds., Chung-Fa chan-cheng 中法战争 (The Sino-French War), Modern Chinese Historical Materials Series, No. 6 (Shanghai: Hsin-chih-shih ch'u-pan-she, 1955), 7 vols., 4,069 pp.

The most important material in these seven volumes is the large selection of previously unpublished documents from the archives of the Grand Council on the Sino-French War (Vol. 5, pp. 499-628; Vol. 6; Vol. 7, pp. 1-118). These represent about one-half of the remainder of the documents on this subject in those archives, and continue the Palace Museum collection Ch'ing Kuang-hsü-ch'ao Chung-Fa chiao-she shih-liao 清光緒朝中法交涉史料 (Documents relating to Sino-French relations in the Kuang-hsü period) which ceased publication in 1937 after reaching August 1884. A generous and very good selection of the important shih-liao documents is reprinted (Vol. 5, pp. 87-498). The collection also contains background material on the war (Vol. 1); materials on Sino-French negotiations and the general situation in southwest China and Tongking, including previously unpublished documents (Vol. 2); descriptions of military operations (Vol. 3); extensive selections from the works of Li Hung-chang and Chang Chih-tung (Vol. 4); and a poor selection of translations from Western-language sources. See The Journal of Asian Studies, 17.1:86-91 (Nov. 1957) for a detailed review. (4,100)

2.9.3 Ti-kuo chu-i yü Chung-kuo hai-kuan, ti-ssu-pien, Chung-kuo hai-kuan yü Chung-Fa chan-cheng 帝国主义与中国海关 第四編中国海关与中法战争 (Imperialism and the Chinese Imperial Maritime Customs, Vol. 4, The Customs and the Sino-French War; Peking: K'o-hsueh ch'u-pan-she, 1957), 248 pp., 1 map.

The series of which this volume is a part is designed to show how the I.M.C. was a tool of foreign imperialism. Its chief value to foreign users lies in the now inaccessible documents from the Customs archives which it prints for the first time. The materials in Vol. 4 consist of: (1) A verbatim translation of H.B. Morse, The International Relations of the Chinese Empire, Vol. II, Chap. 17, "France and Tongking." (Morse is violently denounced as a defender of imperialism whose writings have had a pernicious influence in China. Nevertheless, the editors concede that his account of the Sino-French War is fairly accurate in outline.) (2) Translations of telegrams between Sir Robert Hart and James Duncan Campbell (Chinese Customs Commissioner at London) during the period April 1883 to June 1885. (These have appeared only in part in Stanley Wright, Hart and the Chinese Customs and the French Documents Diplomatiques.) (3) Translations of letters from Hart to Campbell over roughly the same period. (Letters from Campbell to Hart have not yet been translated but are perhaps in preparation.) Both (2) and (3) are from the Customs secret archives. (4) A translation of notes by Campbell on his conference with the French Foreign Secretary, C. de Freycinet in January 1885. (5) Customs Commissioners' reports (previously published in the *Decennial Reports* and the annual *Returns of Trade*). A list of "important personal and geographical names" gives the original in English or French and the Chinese equivalent. See also 2.9.5-6, 2.9.12-13 for the other volumes (Nos. 4-8) thus far published. (9,053)

2.9.4 A Ying 阿英, ed., *Chung-Fa chan-cheng wen-hsüeh chi* 中法战争文學集 (A literary collection on the Sino-French War; Peking: Chung-hua shu-chü, 1957), 24 + 470 pp., illus.

The contents of this volume (the first draft of which was completed in 1937) are grouped under the headings of poetry, novels (extracts), and essays (including memorials and parts of a diary). Among the authors are contemporary officials and military leaders, as well as men of letters. The preface (18 pp.) gives an orthodox interpretation of the war and the Chinese appeasement party, and states that the works are selected for their opposition to imperialism, to Ch'ing corruption, and to the appeasement policy of the Manchu government. Thus patriotism is a dominant theme. Some of the pieces included have only an indirect bearing on the war, presumably because the body of literary writings on the Sino-French conflict itself is rather limited. (9,600)

2.9.5 *Ti-kuo chu-i yü Chung-kuo hai-kuan, ti-wu-pien, Chung-kuo hai-kuan yü Mien Tsang wen-t'i* 帝国主义与中国海关第五編中国海关与緬藏問題 (Imperialism and the Chinese Imperial Maritime Customs, Vol. 5, The Customs and the Burma and Tibet questions;

Peking: K'o-hsueh ch'u-pan-she, 1957), 6 + 204 pp.

As in Vol. 4, the materials here are mainly from the Chinese Imperial Maritime Customs archives: (1) Translations of 110 items of correspondence and telegrams exchanged by Sir Robert Hart and James Duncan Campbell between July 1885 and October 1886 relating to Burma. (These are augmented with pertinent selections from Ch'ing-chi wai-chiao shih-liao and the papers of Chinese officials involved. (2) Translations of 262 telegrams exchanged by Sir Robert Hart and his brother, James Hart, also of the Maritime Customs, between January 1889 and January 1894, relating to the opening of Tibet to trade. (These include relayed messages from the Tsungli Yamen and from the Chinese Resident at Lhasa.) These documents are now published for the first time. Appendix I (pp. 185-200) contains translated selections from the British Parliamentary Papers, Vol. 75, 1885 (as "a reference indicating the British and French efforts to seize colonies in Indochina in this period"). Appendix II is a table of English personal and geographical names and the Chinese equivalents which supplements a similar list in Vol. 4 of this series (item 2.9.3). (2,178)

2.9.6 Ti-kuo chu-i yü Chung-kuo hai-kuan, ti-liu-pien, Chung-kuo hai-kuan yü Chung-P'u Li-ssu-pen ts'ao-yueh 帝国主义与中国海关第六編中国海关与中葡里斯本草約 (Imperialism and the Chinese Imperial Maritime Customs, Vol. 6, The Customs and the Sino-Portuguese Protocol of Lisbon; Peking: K'o-hsueh ch'u-pan-she, 1959), 2 + 98 pp.

This volume contains correspondence and telegrams between Hart and Campbell from the Customs archives dealing with the negotiation of the Protocol of Lisbon in 1886-87, plus a number of communications between Hart and the Tsungli Yamen on the same subject.

The editors' introduction gives a brief and not very lucid summary of some of the issues involved in the Sino-Portuguese negotiations. It refers to Hart's "intrigue" in depriving China of her sovereign rights to Macao, but is on the whole less vituperative about Hart than, for example, the introduction to Vol. 4, item 2.9.3. (3,600)

2.9.7 Li-shih chiao-hsüeh yueh-k'an she 歷史教學月刊社 ("History instruction monthly" society), ed., Chung-Jih chia-wu chan-cheng lun chi 中日甲午战爭論集 (Collected essays on the Sino-Japanese War of 1894; Peking: Wu-shih nien-tai ch'u-pan-she, 1954), 5 + 116 pp.

The titles of the seven articles in this collection reveal something of the doctrinaire approach to this subject currently taken by mainland historians: (1) "The Eastern Learning Society (Tonghak, Tung hsueh tang 東学党)--the anti-feudal and anti-imperialist struggles of Korea (by Chou I-liang 周一良, a Harvard Ph.D.)"; (2) "The discovery of Li Hung-chang's treasonous plot instigated by Timothy Richard before the Shimonoseki Treaty"; (3) "The movement opposing the cession of Taiwan at the time of the Shimonoseki Treaty"; (4) "Hsü Jang 徐驤 and Liu Yung-fu 劉永福 in Taiwan's war of resistance against Japan in 1895 (by Li Kuang-pi 李光璧)"; (5) "The Sino-Japanese War indemnity loans (by Sun Yü-t'ang 孫毓棠)"; (6) "The impact and lesson of American imperialism's aid to Japan in aggressions against China and Korea during the Sino-Japanese War (by Shang Yüeh 尚鉞)"; (7) "Criminal acts of American imperialism in aiding Japan's aggression against China and Korea during the Sino-Japanese War." The exposition of these aspects of the war, says the editor of this volume, will "heighten our political vigilance against imperialism." (8,000)

2.9.8 Chia I-chün 賈逸君 , Chia-wu Chung-Jih chan-cheng 甲午中日战争 (The Sino-Japanese War of 1894-95; Shanghai: Hsin chih-shih ch'u-pan-she, 1955), 2 + 105 pp., 1 map. (12,100)

2.9.9 Cheng Ch'ang-kan 郑昌淦 , Chung-Jih chia-wu chan-cheng 中日甲午战争 (The Sino-Japanese War of 1894; Peking: Chung-kuo ch'ing-nien ch'u-pan-she, 1957), 135 pp.

This small volume is apparently not without some scholarly pretensions,for it cites a variety of primary and secondary sources on the subject; but it is written entirely in a popular vein, with the author telling the story of what happened in vivid, black and white terms. Thus the book becomes a vehicle for the author to praise the virtuous Korean people and to condemn the "traitor" Li Hung-chang, his followers, and his supporters, the American and British imperialists. (25,000)

2.9.10 Shao Hsün-cheng 邵循正 et al., eds., Chung-Jih chan-cheng 中日战争 (The Sino-Japanese War), Modern China's Historical Materials Series, No. 5 (Shanghai: Hsin chih-shih ch'u-pan-she, 1956), 7 vols., 4,209 pp.

About one-third of this collection (Vol. 1, p. 289-Vol. 4, p. 241) consists of documents selected from the Palace Museum publication Ch'ing Kuang-hsü-ch'ao Chung-Jih chiao-she shih-liao 清光緒朝中日交涉史料 (Documents relating to Sino-Japanese negotiations in the Kuang-hsü period). These valuable materials, and selections from other published and previously unpublished Chinese works, together give a fairly comprehensive--although of course, far from complete--coverage of Chinese sources on the Sino-Japanese War of 1894-95. The selection of Japanese and other non-Chinese materials is, however, unbalanced and inadequate. A bibliography, particularly useful for its annotations on Chinese sources, concludes Vol. 7. See the detailed review of this collection by James T. C. Liu in The Journal of Asian Studies, 17.1:92-98 (Nov. 1957). (5,000)

2.9.11 A Ying 阿英, ed., Chia-wu Chung-Jih chan-cheng wen-hsueh chi 甲午中日战争文學集 (A literary collection on the Sino-Japanese War of 1894-95; Peking: Chung-hua shu-chü, 1958), 563 pp., 9 plates.

The chief merit of A Ying's literary collections lies undoubtedly in the information they provide on the historical periods under consideration, however oblique and diffuse that information may be. Like the other collections, this one pursues the central theme of Chinese patriotism (indignation at the national humiliation, dissatisfaction with the Ch'ing government, etc.) as reflected in writings of these categories: (1) poetry, (2) novels, (3) "war diaries," and (4) essays. Among the poets, the best known is Huang Tsun-hsien 黄遵憲 (1848-1905); many obscure poets are included whom the editor does not identify. Among the novels, Tseng P'u's 曾樸 Nieh hai hua 孽海花 is included in abridged form. Under essays are some memorials to the Ch'ing court (e.g., by Chang Chih-tung). (5,100)

2.9.12 Ti-kuo chu-i yü Chung-kuo hai-kuan, ti-ch'i-pien, Chung-kuo hai-kuan yü Chung-Jih chan-cheng 帝国主义与中国海关第七編中国海关与中日战争 (Imperialism and the Chinese Imperial Maritime Customs, Vol. 7, The Customs and the Sino-Japanese War; Peking: K'o-hsueh ch'u-pan-she, 1957), 244 pp.

The materials from the Customs archives translated for this volume include: (1) Letters from the Korean Customs Chief Commissioner, H. F. Merrill, to Sir Robert Hart, covering the period 1885-89. Merrill and Morse both graduated in 1874 from Harvard, where their letter-books are deposited. (2) 554 letters and telegrams exchanged between Hart and James Duncan Campbell from May 1894 through April 1896 which treat, among other things, the purchase of foreign warships and the enlistment of foreign officers, securing wartime loans, and the post-war scramble over the indemnity loans to China. For this group of documents the editors have not supplied their usual explanatory remarks. (3) Reports from the Customs Commissioner at Tamsui, H. B. Morse, to Robert Hart between February and June 1895 concerning conditions on Taiwan. This volume is the only one in the series (Vols. 4-8 published to date) that contains no prefatory remarks by the editors nor any interpretive material aside from the chapter headings. (1,745)

2.9.13 *Ti-kuo chu-i yü Chung-kuo hai-kuan, ti-pa-pien, Chung-kuo hai-kuan yü Ying Te hsü-chieh-k'uan* 帝国主义与中国海关第八編中国海关与英德續借款 (Imperialism and the Chinese Imperial Maritime Customs, Vol. 8, The Customs and the additional loans from Britain and Germany; Peking: K'o-hsueh ch'u-pan-she, 1959), 60 pp.

This volume contains translations of 159 letters and telegrams exchanged between Robert Hart and James Duncan Campbell during the period May 1896-April 1899, dealing with the negotiations for the Anglo-German loans of 1896 and 1898 which were secured on the Customs revenue. The editors' preface states that although Hart maintained "a surface calm" while the powers fought to place loans with the Chinese government, he was in fact instrumental in aiding the British cause. An appendix contains two memorials from the Hu-pu (Board of Revenue), dated June 11, 1896, concerning the Anglo-German loans; a table showing foreign loans, payments, and interest rates for 1887-98; and exchanges between Hart and the Tsungli Yamen about likin. The documents from the Customs archives are printed here for the first time. (4,000)

2.9.14 Institute of Historical Research, Chinese Academy of Sciences, ed., Hsi-liang i-kao, tsou-kao 錫良遺稿奏稿 (Posthumous papers and memorials of Hsi-liang, Peking: Chung-hua shu-chü, 1959), 2 vols., 56 + 1344 pp.

These two volumes contain 1215 memorials by this Mongol bannerman official. In the years between 1900 and 1911, Hsi-liang (1853-1917) served as governor or governor-general of a succession of provinces including Shanshi, Honan, Szechwan, and Manchuria. His memorials, previously unpublished, throw light on many important developments in late Ch'ing history, e.g., the Szechwan-Hankow Railway, Japanese intentions toward Manchuria after the Russo-Japanese War, American loan negotiations for railway construction, the Manchu reform movement, and negotiations with Western powers. The introduction gives a chronology of Hsi-liang's official career. The memorials (which begin on April 9, 1898 and end on January 31, 1912) are listed by title in the tables of contents of the two volumes. Publication of Hsi-liang's official dispatches (cha-tu 札牘) and telegrams (tien-kao 電稿) is promised in due course. (1,300)

2.9.15 Ch'ing Ju-chi 卿汝楫, Mei-kuo ch'in Hua shih 美国侵华史 (A history of American aggression against China; Peking: San-lien shu-tien, 1932 and 1956), 2 vols., Vol. 1 (1952) 11 + 266 pp.; Vol. 2 (1956) 10 + 630 pp.

These volumes consider Sino-American relations from 1874 to 1864, and from 1864 to 1900 respectively. Much detailed material from both Chinese and American archival and other sources is brought forth in order to illustrate the single theme of American aggression, as manifested in the spheres of trade, diplomacy, war, and the like. All this is narrated in a highly colored language. The author's preface states that the "so-called authoritative" Western works on Sino-American relations are all designed as "opiates to drug the Chinese people." Each volume has an appendix containing a sizable list of publications cited in the text. (6,000)

2.9.16 Shao Hsi 紹溪, Shih-chiu shih-chi Mei-kuo tui Hua ya-p'ien ch'in-lueh 十九世紀美国对华鴉片侵略 (American aggression against China with regard to opium in the nineteenth century; Peking: San-lien shu-tien, 1952), 3 + 110 pp. (10,000)

2.9.17 Chu Shih-chia 朱士嘉, comp., Shih-chiu shih-chi Mei-kuo ch'in Hua tang-an shih-liao hsüan-chi 十九世紀美国侵华檔案史料选輯 (Selected archival materials on American aggressions against China in the nineteenth century; Peking: Chung-hua shu-chü, 1959), 2 vols., 4 + 499 pp.

Documentary materials on Sino-American relations between 1836 and the Boxer Rebellion, principally in the form of diplomatic communications (chao-hui 照會), as well as a number of memorials, letters, and proclamations. Some of these codument s were obtained from the Chinese archives, and many of the rest from the U.S. National Archives. These last were copied by the compiler in 1940, and consist largely of attachments to reports from the U.S. Minister at Peking or to consular reports from Canton, Shanghai, and other places. Most of these documents were originally in Chinese; a few items are translated into Chinese from the English. The compiler has arranged his selections under ten tendentiously-worded headings (e.g., "the intrigue of the American aggressors during the Taiping Revolution," "America's economic plunder of China in the nineteenth century," and "America's cultural aggression by means of religion"). (3,100)

2.9.18 Chu Shih-chia 朱世嘉, ed., Mei-kuo p'o-hai Hua-kung shih-liao 美国迫害华工史料 (Historical materials concerning America's persecution of Chinese laborers; Peking: Chung-hua shu-chü, 1958), 164 pp.

The editor (a Columbia Ph.D.) in his preface tells of American aggression against China from the 1860's on, which allegedly consisted on the one hand of plundering of China's resources, and on the other, of luring Chinese workers to America where they were subjected to inhumane treatment. This volume, intended to provide documentary material illustrating the latter, is divided into four sections: (1) "American criminal acts of kidnapping and defrauding Chinese laborers" (documents, according to the author, largely from the U.S. National Archives); (2) "American criminal acts of persecuting Chinese laborers" (mainly from the Tsungli Yamen archives);

(3) "The attitude of the Ch'ing government toward American persecution of Chinese laborers" (documents from the Ch'ing-chi wai-chiao shih-liao 清季外交史料); and (4) "The Chinese people's boycott of American goods" (largely from contemporary Chinese newspapers). The materials from the U.S. Archives consist of exchanges between Chinese and American officials and of excerpts from U.S. consular reports. Presumably they are here translated from English into Chinese, but there is no statement by the editor that this is the case. Nor is there any indication of the precise origin of the individual documents. Some of the documents reproduced here substantiate the evils of the "coolie trade." Others are neutral (e.g., contract forms) and do not necessarily support the tendentious title or subtitles of the book. While there is clearly little to be proud of in the American record in this matter, the present collection is extremely selective and is far from providing all the materials necessary for a full study of the subject. An anti-American literary collection on the exclusion of Chinese laborers, Fan Mei Hua-kung chin yueh wen-hsueh chi, edited by A. Ying, has been announced, but so far it has not been seen. (2,100)

3. THE REPUBLIC

This section includes works which deal with the period from the end of the Manchu Dynasty until the establishment of the People's Republic of China in October 1949. It is obvious that the closer we come to the present, the more difficult it is for the mainland historians to be dispassionate and objective in their treatment of historical events and personalities. In general, the volumes in Section 3, especially in sec. 3.3, are more visibly propagandistic than the works discussed in other parts of this bibliography. The publication of source materials continues to be an important part of current output, but perhaps to a lesser extent than for the nineteenth century. As we have pointed out in the introduction to this volume, the perils involved in dealing with "contemporary history" are many, which perhaps accounts for the relatively smaller number of monographic treatments of the period after 1919. Given the current orientation of the Party's historical line toward "emphasizing the present and de-emphasizing the past" (*hou-chin po-ku*), however, it may be predicted that the next few years will see increasing attention to the years covered by Section 3. Additional materials on the Republican period will also be found in Sections 4 (economic history) and 5 (intellectual and cultural history).

3.1 THE 1911 REVOLUTION AND THE WARLORD PERIOD

To the mainland historians 1911 was a "bourgeois revolution" that failed. It was the culmination of the "Old Democratic Revolution" which after 1919 gave way to the "New Democratic Revolution" under the leadership of the Chinese Communist Party. Sun Yat-sen and his Republican movement are treated with due respect in mainland China, but an attempt is also made to demonstrate the popular nature of the revolution which extended beyond the confines of the T'ung-meng Hui and its affiliated societies. The overthrow of the Manchu Dynasty was a positive and progressive act, but the revolutionary cause was betrayed by the warlords and the constitutionalists. Implicitly, Sun Yat-sen is criticized for placing his trust in anything other than the popular movement and the military force which it might generate. Nevertheless, because of his friendly relations with the Soviet Union after 1917, his policy of cooperating with the Chinese Communist Party, and also the importance of Sun's popular image to the Chinese nationalism which the Communists seek to harness, he has remained an honored member of the revolutionary pantheon.

The first five items below are general accounts of the 1911 Revolution. They are followed by four titles dealing with Sun Yat-sen and other revolutionary leaders. Perhaps the most valuable contribution being made by the mainland historians to the study of the 1911 Revolution is their extensive publication of source materials (3.1.10-18). Yuan Shih-k'ai as president, and the warlord regimes which followed have received relatively less attention; the last three items in this section are concerned with these subjects.

3.1.1 Kao Shao-hsien 高韶先, <u>Hsin-hai ko-ming</u> 辛亥革命 (The 1911 Revolution; Shanghai: Shih-tai shu-chü, 1950), 92 pp. (?)

3.1.2 Ch'en Hsü-lu 陳旭麓, <u>Hsin-hai ko-ming</u> 辛亥革命 (The Revolution of 1911; Shanghai: Shang-hai jen-min ch'u-pan-she, 1955), 2 + 129 pp.

> Intended as a reference work for college students, this book summarizes the history of the years between Sun Yat-sen's formation of the T'ung-meng Hui 同盟會 (1905) and Yuan Shih-k'ai's usurpation of the presidency (1912). The account is quite detailed, but it carries a heavy burden of clichés

and unimaginative interpretation. (26,000)

3.1.3 Li Shih-yueh 李時岳, Hsin-hai ko-ming shih-ch'i liang-Hu ti-ch'ü ti ko-ming yun-tung 辛亥革命時期兩湖地區的革命運動 (The revolutionary movements in Hupei and Hunan in the period of the 1911 Revolution; Peking: San-lien shu-tien, 1957), 132 pp.

This volume focuses attention on the provinces of Hupei and Hunan in the 1911 Revolution and its aftermath, as the area where revolutionary activities first took shape and where the revolutionary cause was soon betrayed by the warlords and constitutionalists. In addition to considering the revolutionary societies led by "bourgeois intellectuals," this study also briefly discusses the "struggles" of the peasants and laborers in the same period. The author contends that, after they were robbed of the "fruits of the revolution," the rank-and-file revolutionists rose up in revolt. His development of this theme is limited by the lack of sources, however, the only available ones being scattered newspaper items. (15,000)

3.1.4 Ts'ai Chi-ou 蔡寄鷗, O-chou hsueh-shih 鄂州血史 (A history of bloodshed in Hupei; Shanghai: Lung-men lien-ho shu-chü, 1958), 11 + 257 pp.

The author was a newspaper writer and editor during the 1911 Revolution and was on the scene when the Wuchang uprising took place. This book contains detailed accounts of revolutionary activities, primarily in Hupei, and also describes major developments up to the dissolution of parliament by Yuan Shih-k'ai in 1913. Unfortunately, the author's attempt at vivid writing or fictional style--which he claims does not interfere with historical fact--leaves one in doubt as to the exact boundary between the two. In many places, moreover, he conjectures from controversial evidence, as he admits. This is a posthumous publication, the manuscript having been edited by the Third Office of the Institute of Historical Research, Chinese Academy of Science, which is responsible for occasional footnotes pointing out discrepancies with other sources. A bibliography is appended. (2,640)

3.1.5 Yang Yü-ju 楊玉如, Hsin-hai ko-ming hsien-chu chi 辛亥革命先著記 (An account of the beginnings of the 1911 Revolution; Peking: K'o-hsueh ch'u-pan-she, 1958, preface 1952, colophon 1954), 8 + 285 pp.

In the view of the author, the Wuchang uprising, in which he appears to have taken part, marked the decisive beginning of the 1911 Revolution. The narrative deals with background developments from 1908 on and is rich in details of the organization and activities of the revolutionary groups at Wuchang, the establishment of the Hupei Military Government and its battles against the Ch'ing forces, and the complex events attendant on the origins of the Republic of China. The numerous documents quoted include official proclamations, personnel lists, and military orders. This is a Hupei-centric reference work which, unfortunately, has no bibliography and gives no indication of the provenance of its valuable documentation. (8,174)

3.1.6 Sun Chung-shan hsuan-chi 孫中山选集 (Collected works of Sun Yat-sen; Peking: Jen-min ch'u-pan-she, 1956), 2 vols., 922 pp., 21 plates, map.

Vol. 1 includes selections from Sun's writings prior to the reorganization of the Kuomintang in 1923; Vol. 2, materials from the last two years of Sun's life. The publisher asserts that because a good edition of Sun's works has never been published, the correct text of some items included herein has been difficult to establish. The emphasis in these volumes on the period of Sun's alliance with the CCP, and the choice of selections, seem obviously designed to stress Sun's connections with communism. (10,000)

3.1.7 Ch'en Hsi-ch'i 陳錫祺, T'ung-meng Hui ch'eng-li ch'ien ti Sun Chung-shan 同盟會成立前的孫中山 (Sun Yat-sen before the establishment of the T'ung-meng Hui; Canton: Kuang-tung jen-min ch'u-pan-she, 1957), 84 pp., 4 plates. (10,120)

3.1.8 Ho Hsiang-ning 何香凝, Hui-i Sun Chung-shan ho Liao Chung-k'ai 回忆孫中山和廖仲愷 (Reminiscences about Sun Yat-sen and Liao Chung-k'ai; Peking: Chung-kuo ch'ing-nien ch'u-pan-she, 1957), 40 pp.

Liao (1878-1925), a leader of the left wing of the Kuomintang, was assassinated in Canton shortly after Sun Yat-sen's death. This marked the beginning of open dissension between the radical and conservative factions of the KMT. Ho Hsiang-ning was Liao's wife. (25,000)

3.1.9 Shanghai Editorial Office of Chung-hua shu-chü, ed., Ch'iu Chin shih chi 秋瑾史跡 (Historical evidences of Ch'iu Chin [a female martyr in the 1911 revolutionary movement]; Peking: Chung-hua shu-chü, 1958), 11 + 234 pp.

This volume contains photographs, autographs, and the like. (2,000)

3.1.10 Ch'ai Te-keng 柴德賡 et al., eds., Hsin-hai ko-ming 辛亥革命 (The 1911 Revolution), Modern Chinese Historical Materials Series, No. 10 (Shanghai: Shang-hai jen-min ch'u-pan-she, 1957), 8 vols., 4,489 pp. + illus.

This is the last of the Modern Chinese Historical Materials Series to have been published at the time of this writing, although three other titles are promised. The present collection is divided chronologically into four parts, which together treat the background and course of the revolution which brought the Ch'ing dynasty to an end and established the Republic of China. Part I (Vol. 1) and Part II (Vols. 2-4) are devoted respectively to the revolutionary movement during the periods of the Hsing-Chung Hui 興中會 and the T'ung-meng Hui. Materials on the founding of these societies, on the numerous abortive uprisings prior to 1911, and on the Manchu constitutional movement are reprinted from the accounts of participants or contemporaries; in addition, a large amount of hitherto unpublished documentation from the archives of the Manchu government is made available. Part III deals with the Wuchang uprising of October 1911 and with the subsequent spread of the revolution through the provinces (Vols. 5-7). Part IV is devoted to the Provisional Government at Nanking, to the peace negotiations between North and South China, and to the establishment of the Republic of China. The editors have drawn their material from 117 works which they list with annotations at the end of Vol. 8; they also supply an annotated list of 133 other works which they used for reference. (3,550)

3.1.11 Min Pao 民報 (Peking: K'o-hsueh ch'u-pan-she, 1957), 4 vols. + supplement (1958).

A complete file of the principal T'ung-meng Hui organ, Min Pao (1905-08, 1910), has been assembled by the Third Office of the Institute of Historical Research in cooperation with the Central Museum of Revolution, and reproduced in its entirety by photo-offset. The twenty-six issues, including illustrations and advertisements as well as all the articles, are arranged in four thick volumes. In addition, the "summer supplement" to No. 15, which subsequently became available, is reproduced in a separate thin volume. Apart from some brief remarks about the historical siginificance of Min Pao (it countered the spreading influence of the "reformists"), and an explanation of the various editions of the newspaper, the editors have supplied no introduction. (2,655)

3.1.12 Hupei Committee of the Chinese People's Political Consultative Conference, ed., Hsin-hai shou-i hui-i lu 辛亥首义回忆錄 (Recollections on the first rising of the 1911 Revolution; Wuhan: Hu-pei jen-min ch'u-pan-she, 1957), 2 vols., 6 + 218 + 226 pp., illus., 1 map.

The Chinese Communist regime's interest in the "people's revolutions" and in "setting straight" and amplifying the record of these revolutions has prompted the Hupei Committee of the CPPCC to collect through interviews or in writing the reminiscences of surviving participants in the Wuchang rising which set off the 1911 Revolution. A few of the twenty-odd contributors, all men now in their sixties or seventies, were members of the T'ung-meng Hui. A larger number belonged to lesser revolutionary groups (such as the Jih-chih Hui 日知會 and Kung-chin Hui 共進會) of which accounts are also given. To each of the individual statements, which vary considerably in length and in style, the editors have added a brief biography of the contributor and extensive explanatory notes. This is a good example of the type of "oral history" project that is increasingly popular in mainland China. (17,000)

3.1.13 Chang Kuo-kan 張國淦, Hsin-hai ko-ming shih-liao 辛亥革命史料 (Historical materials on the 1911 Revolution; Shanghai: Lung-men lien-ho shu-chü, 1958), ii + 338 pp.

These materials on the 1911 Revolution are arranged under four main headings: (1) the Wuchang uprising; (2) uprisings in other provinces in response to Wuchang; (3) the North-South negotiations; and (4) the abdication of the Ch'ing emperor. A native of Hupei and a participant in the revolutionary movement there, the author has collected more material on that province than any other (e.g., membership lists of secret revolutionary societies in existence in Hupei before the Wuchang rising). The text is in the form of a chronological day-by-day narrative summary by the author which is interspersed with frequent and lengthy quotations from documents. Sources are not always clearly indicated. Pages 323-338 provide a chronological chart of major events in the revolution. (6,537)

3.1.14 Chou Shan-p'ei 周善培, Hsin-hai Ssu-ch'uan cheng-lu ch'in-li chi 辛亥四川爭路親歷記 (A record of my personal experiences in the Szechwan railway strife of 1911; Chungking: Ch'ung-ch'ing jen-min ch'u-pan-she, 1957), 65 pp.

Chou Shan-p'ei was an "industrial taotai" in charge of railroad construction in Szechwan province, at the time of the uprising precipitated by the nationalization of the Szechwan-

Hankow and Szechwan-Canton railways which led to the fall of the Ch'ing dynasty. At the age of eighty, he has written his memoir on that incident, which lasted roughly from May to October 1911. Interspersed in his narrative are some documents, the most interesting of which is probably a memorial drafted by himself and submitted to the throne by the acting governor-general Wang Jen-wen 王人文 in protest against the Manchu government's railroad nationalization plan. The local officials and gentry who were involved in the affair are identified in two tables (pp. 5-6). Appended are two short essays the author worte to commemorate Wang Jen-wen (1922) and Ch'en Ch'ung-chi 陳崇基, a member of the Szechwan gentry and legal expert (1933). (4,500)

3.1.15 Tai Chih-li 戴執禮 ed., Ssu-ch'uan pao-lu yun-tung shih-liao 四川保路運動史料 (Historical materials on the Szechwan "Railway Protection Movement"; Peking: K'o-hsueh ch'u-pan-she, 1959), 42 + 548 pp.

This is an impressive collection of documents on the Szechwan Railway Protection Movement of 1911, quite a number of which were hitherto unpublished or hard to find. Memorials, proclamations, correspondence, handbills, diaries, and the like, totaling 475 items, are arranged in five chronological groups centering around these events: (1) internal and external developments related to the Szechwan-Hankow Railway (July 1903-May 1911); (2) the Ch'ing government's measures to nationalize the railway and its clash with the "constitutionalists" (May 5-June 17, 1911); (3) the formation of the Railway Protection League and Szechwan's demand for independence (June 17-Sept. 6, 1911); (4) armed revolt of the Szechwanese and the Ch'ing surrender (Sept. 7-Nov. 21, 1911); and (5) the Chungking military government and its dissolution (Nov. 22, 1911-May 8, 1912). The editor's preface discusses the importance of the movement and the reasons for its ultimate failure as Communist historians see them. Readability of the documents has been improved by the addition of punctuation and minor deletions and changes. A bibliography (pp. 541-548) is appended, and photographs of a few mementos of the movement have been included. (1,780)

3.1.16 Third Office, Institute of Historical Research, Chinese Academy of Science, ed., Yun-nan Kuei-chou hsin-hai ko-ming tzu-liao 云南貴州辛亥革命資料 (Source materials on the 1911 Revolution in Yunnan and Kweichow; Peking: K'o-hsueh ch'u-pan-she, 1959), 4 +317 pp., illus.

This collection assembles miscellaneous materials on Yunnan and Kweichow in the 1911 Revolution which are not included in item 3.1.10. Among the primary materials are such documents as "A warning to Yunnan" written by Yunnan students in Indochina (ca. 1906); declarations by the T'ung-meng Hui from Japan against foreign loans, "bogus" constitutional government, etc. (1911); and "A letter of blood and tears from Kweichow." Also included are diaries and personal recollections of a number of individuals. In addition there are brief biographies of "martyrs" from the two provinces, collected from both published and unpublished sources. The editors point out that there exist many inconsistencies among the records reproduced here, as well as discrepancies with other records of the revolution. There is a brief explanatory note before each item. (2,700)

3.1.17 Tsou Jung 鄒容, Ko-ming chün 革命軍 (The Revolutionary Army; Shanghai: Chung-hua shu-chü, 1958, first published 1903), 44 pp., 2 plates.

Tsou's (1885-1905) book was an outspoken attack on the Ch'ing dynasty, largely from a nationalistic, anti-Manchu position. (7,000)

3.1.18 Chang Nan 張枬 and Wang Jen-chih 王忍之, comps., Hsin-hai ko-ming ch'ien shih-nien chien shih-lun hsuan-chi 辛亥革命前十年间時論选集 (Selected articles from periodicals and books for the ten years before the 1911 Revolution; Peking: San-lien shu-tien, 1960), Part 1, 2 vols., 26 + 970 pp., plates.

Selections from eighteen journals and seven books published during the years 1901-04. Parts 2 and 3 (not yet published) will continue this collection through 1911. Each source is carefully identifed and described on pp. 966-970. The aim of the selections is to provide materials for the study of the social ideology of the "national bourgeois class" (min-tsu tzu-ch'an chieh-chi 民族資產階級) which "led" the revolution that overthrew the Manchu dynasty. (5,000)

3.1.19 Ch'en Po-ta 陳伯達, Ch'ieh-kuo ta-tao Yuan Shih-k'ai 竊国大盜袁世凱 (Yuan Shih-k'ai, the traitorous thief; Peking: Jen-min ch'u-pan-she, 1949, first ed., Chungking 1945), 74 pp.

Like Ch'en Po-ta's book on Chiang Kai-shek (3.3.1), this small volume is strong on polemics. The author states in his foreword to this edition that he wrote the book in 1945 to satirize Chiang Kai-shek, i.e., "some of the major anti-revolutionary techniques of Yuan described here also apply to Chiang." This is spelled out at the end by such expressions as "Yuan Shih-k'ai has come back to life in new clothes." For his material on Yuan's usurpation of power and other "treacheries," the author cites half-a-dozen secondary works in a postscript; the text is by and large undocumented. (65,000)

3.1.20 Li Shu 黎澍, Hsin-hai ko-ming ch'ien-hou ti Chung-kuo cheng-chih 辛亥革命前後的中国政治 (Chinese politics before and after the 1911 Revolution; Peking: Jen-min ch'u-pan-she, 1954), 3 + 152 pp.

This is a revised version of the author's earlier Hsin-hai ko-ming yü Yuan Shih-k'ai 辛亥革命与袁世凱 (not seen). The new title is rather misleading, for the part before the revolution receives a cursory treatment. By comparison, the machinations of Yuan Shih-k'ai after the revolution are discussed in greater detail, but in neither case are the author's sources indicated, except for quotations from Marx and Lenin. In the period after Yuan's death in 1916 and before the May Fourth Movement in 1919, the two major developments are described as: (1) the growth of the Chinese proletariat and national industry (for the latter a smattering of statistics is given and the indirect cause is said to be the Western powers' preoccupation with World War I); and (2) the inspiration provided by the Russian October Revolution for the development of Communism in China. (80,000)

3.1.21 T'ao Chü-yin 陶菊隱, Pei-yang chün-fa t'ung-chih shih-ch'i shih-hua 北洋軍閥統治時期史話 (A historical narrative of the Warlord Period; Peking: San-lien shu-tien, 1958), 218 + 240 + 232 + 201 + 230 pp.

Volumes 1-5 of this publication (of which there presumably will be more) cover a period of twenty-five years between 1895 and 1920, from the emergence of the Peiyang warlords to the end of the Chihli-Anhwei civil war. The minutely detailed descriptions of the events in this period suggest that the author was intimately acquainted with the political figures involved. But there is in general no indication of how and where he obtained his information. The forty-nine chapters are divided into several sections each, and all have headings the style of which is somewhat reminiscent of the traditional Chinese novel. The narrative is occasionally interlarded with bits of dialogue, quotations from letters and the like. Needless to say, the usefulness of these materials is impaired by the mystery of their origin, but explanatory footnotes at the end of each chapter identify people and obscure references. (19,500)

3.2 THE MAY FOURTH MOVEMENT

The Chinese Communist interpretation of the May Fourth Movement tends to stress its patriotic, anti-imperialist character and the entrance of the proletariat and its vanguard, the Chinese Communist Party, onto the political stage in modern China. While the intellectual efflorescence of the years immediately following World War I, with its many cross currents, is not ignored, those streams which seem to lead directly to the Chinese Communist movement are stressed disproportionately. At the same time "liberal" thought, as represented for example by Hu Shih and Liang Shu-ming, is condemned as diversionary and the dupe of the imperialist aggressors. The bourgeoisie had had its chance, but the leadership of the revolution after 1919 (now called the New Democratic Revolution) falls inevitably to the working class and the Communist Party. The first four items in this section are brief histories of the May Fourth Movement arranged in order of their publication. Items 3.2.5 and 3.2.6 are collections of essays on the May Fourth period which treat in greater detail most of the matters discussed in the survey treatments. The next three titles are valuable collections of source materials. In this connection, the reader of this section will probably be interested in

Chow Ts'e-tsung, Research Guide to the May Fourth Movement (Cambridge: Harvard University Press, 1961).

3.2.1 Hua Kang 華崗, Wu-ssu yun-tung shih 五四運動史 (A history of the May Fourth Movement; Shanghai: Hsin wen-i ch'u-pan-she, 1951), 3 + 226 pp.

This is a more elaborate version of the chapter on the May Fourth Movement in Hua Kang's Chung-kuo min-tsu chieh-fang yun-tung shih (1.4.9). It has been issued by several other Chinese publishers since 1951, and in 1952 was translated into Japanese by Amano Motonosuke 天野元之助 and others. A long concluding chapter (pp. 179-226) offers a detailed Marxist-Leninist-Maoist treatment of the "historical meaning and lessons" of May Fourth. (34,000)

3.2.2 Chia I-chün 賈逸君, Wu-ssu yun-tung chien-shih 五四运动簡史 (A brief history of the May Fourth Movement; second ed., Shanghai: Shih-hsi ch'u-pan-she, 1953), 43 pp. (first ed., Peking, 1951).

The May Fourth incident and some of the intellectual strains and political issues that led up to it are set forth here in a cursory fashion, while for the "proper" evaluation of the movement the author turns to Mao Tse-tung. Texts of manifestos of the Peking and Shanghai students, and the Chinese delegation's protest to the Paris Peace Conference are appended. (5,000)

3.2.3 Hung Huan-ch'un 洪焕春, Wu-ssu shih-ch'i ti Chung-kuo ko-ming yun-tung 五四時期的中国革命运动 (The Chinese revolutionary movement during the May Fourth period; Peking: San-lien shu-tien,

1956), 2 + 187 pp.

After an introductory chapter on the economy and politics in China around the turn of the century, this volume discusses the "early phase of the new culture movement" (ca. 1915-19) the May Fourth Movement (1919), and the spread of Marxism-Leninism up to the founding of the Chinese Communist Party in 1921. Among the sources are newspapers and periodicals of the period, and a number of manifestos and speeches are quoted at considerable length. Some space is devoted to the literature of the May Fourth period--mostly with reference to Lu Hsün 鲁迅, and also to Hu Shih 胡適 and Liang Shu-ming 梁漱溟, both of whom are condemned. The account of the founding of the CCP (pp. 161-174) plays down the role of Ch'en Tu-hsiu 陳獨秀, while it is replete with the activities of Mao Tse-tung. A fully orthodox account, it adds little that is new. (10,000)

3.2.4 Ting Shou-ho 丁守和, Yin Hsü-i 殷叙彝, and Chang Po-chao 張伯昭, Shih-yueh ko-ming tui Chung-kuo ko-ming ti ying-hsiang 十月革命对中国革命的影響 (The impact of the Russian October Revolution on China; Peking: Jen-min ch'u-pan-she, 1957), 2 + 180 pp. (14,000)

3.2.5 Division of Marxist-Leninist Materials, Central China Institute of Engineering, ed., Wu-ssu yun-tung wen-chi 五四运动文輯 (Selected essays on the May Fourth Movement; Wuhan: Hu-pei jen-min ch'u-pan-she, 1957), 175 pp.

The twelve essays that begin this volume are reprinted from mainland Chinese periodicals and newspapers of the last decade. Those in Part I (pp. 1-66) discuss the May Fourth Movement in relation to the Russian October Revolution, "sabotage" by the imperialists, the workers, the "June 3" incident, Li Ta-chao 李大釗, and Hu Shih. Part II (pp. 67-98), built around the theme of "ideological class struggles," contains articles on the spread of Marxism in China, Li Ta-chao's ideology, the "Socialist Debate," and the Peiyang warlords' destruction of progressive periodicals. An appendix (pp. 99-175), about the length of Parts I and II combined, reprints from the February 1955 number of the journal Chin-tai-shih tzu-liao 近代史資料 (Source materials on modern history) some documents of the Society for the Study of Marxism at Peking University (1921-22), three chapters from a book on May Fourth by Ts'ai Hsiao-chou 蔡曉舟 and Yang Ching-kung 楊景工 published in 1919, some articles in contemporary newspapers and magazines, and finally two reports on the May Fourth demonstrations written by a lieutenant in the Peking garrison. (18,000)

3.2.6 Shanghai Philosophical Society and the Social Sciences Society, ed., Chi-nien wu-ssu yun-tung ssu-shih chou-nien lun-wen chi 紀念五四运动四十週年論文集 (Collected essays commemorating the fortieth anniversary of the May Fourth Movement; Shanghai: Shang-hai jen-min ch'u-pan-she, 1959), 3 + 193 pp.

Eleven articles on various facets of May Fourth, including the Shanghai labor strikes, the intellectuals, the new culture and new literature movements, the press and periodicals, Li Ta-chao, Ch'ü Ch'iu-pai 瞿秋白, Lu Hsün, the "science and metaphysics debate," and Hu Shih. In general, these articles repeat or elaborate the official Party view of the May Fourth movement, and contribute little that is new in either material or interpretation. (16,000)

3.2.7 Research Section of the Bureau of Translation of the Works of Marx, Engels, Lenin, and Stalin, Central Committee of the CCP, comp., Wu-ssu shih-ch'i ch'i-k'an chieh-shao 五四時期期刊介紹 (An introduction to periodicals of the May Fourth period; Peking: Jen-min ch'u-pan-she), 3 vols., Vol. 1, 1958, 5 + 833 pp., illus.; Vol. 2, 1959, 10 + 962 pp., illus.; Vol. 3, 1959, 10 + 1121 pp., illus.

These three thick volumes cover a total of 160-odd periodicals which were published in China roughly between 1919 and 1923. Each volume contains first a section in which the periodicals are described in some detail as to their auspices, contributors, ideological position, influence on the public, and the like. This section is followed by another which contains manifestoes of some of the periodicals described, indices to articles by subject, and/or lists of articles by issue for individual periodicals. Needless to say, this is a very useful compilation, even though both the selection of periodicals and much of the "content analysis" show a dominant concern with political ideology to the exclusion of other (e.g., literary) considerations. (3,000)

3.2.8 Modern Historical Materials Compilation Section, Third Office, Institute of Historical Research, Chinese Academy of Sciences, ed., Wu-ssu ai-kuo yun-tung tzu-liao 五四爱国运动資料 (Materials on the May Fourth patriotic movement; Peking: K'o-hsueh ch'u-pan-she, 1959), 5 + 865 pp. + illus.

This compilation incorporates: (1) seven books originally published in 1919; (2) a selection from the archives of the Shanghai International Settlement Police Department (pp. 709-786); and, (3) "A daily record of the patriotic movement from May 4 to

June 30," apparently hitherto unpublished and copied for the present volume from newspaper accounts. The book titles are Ch'ing-tao ch'ao 青島潮 (The tumult of Tsing-tao); Hsueh-chieh feng-ch'ao chi 学界風潮紀 (An account of school strikes); Shang-hai pa-shih shih-lu 上海罷市實錄 (A true record of the Shanghai strike); Min-ch'ao ch'i-jih chi 民潮七日記 (Seven days of the people's turbulence); Shang-hai pa-shih chiu-wang shih 上海罷市救亡史 (A history of the Shanghai strike and rescue from destruction); Chang Tsung-hsiang 張宗祥; and Ts'ao Ju-lin 曹汝霖. These books themselves are largely compilations which rely heavily on current newspapers. Of the original compilers, nothing is known, not even their real names in some instances. The editors of the present volume complain that they were not only uncritical as compilers but were also deceived by British and American imperialists (because the latter are sometimes represented as being sympathetic to the May Fourth Movement). Nonetheless, the editors have left this kind of material intact. The translation of the International Settlement police archive was prepared for this volume by the Shanghai Municipal Archives. The direct relevance of this material to the May Fourth Movement seems questionable, although the editors claim that it reveals the imperialists' repression of the "people's anti-imperialist movement," as well as the alignment of class forces in Shanghai in 1919. (10,500)

3.2.9 Institute of Historical Research, Shanghai Academy of Social Sciences, comp., Wu-ssu yung-tung tsai Shang-hai shih-liao hsuan-chi 五四運動在上海史料选輯 (Collected historical materials on the May Fourth Movement in Shanghai: Shanghai: Shang-hai jen-min ch'u-pan-she, 1960), 36 + 804 pp., 26 plates.

An extensive collection of items, mainly from Shanghai newspapers for the year 1919, which report on events primarily in Shanghai during the May Fourth Movement. There is, for example, a considerable amount of material on the strikes by students and workers. Under the general heading "The outbreak of the May Fourth Movement and its historical conditions and causes" appear a series of statistical tables on miscellaneous aspects of the Shanghai industries, labor force, and the like, for that period. The section headings are worded to highlight such themes as the influence of the Russian October Revolution, the tremendous public sentiment of patriotism, and "British and American intrigue to wreck the May Fourth Movement" (even though the selections from reports of the American consul in Shanghai during this period do not bear out that title). The editors' interpretive remarks which precede each section of source materials are rather minimal. (5,000)

3.2.10 Wu-ssu yun-tung (hua-ts'e) 五四运动画册 (The May Fourth Movement--a pictorial volume; Peking: Wen-wu ch'u-pan-she, 1959), n.p.

This volume consists of reproductions of photographs accompanied by captions of various lengths (e.g., Mao Tse-tung as a young man, Li Ta-chao, handbills and publications of the May Fourth period, scenes of student parades, and one picture each of Marx and Lenin). Some of these reproductions are quite blurred. A nine-page text at the back contains a brief account of events in the May Fourth period, with emphasis on the impact of Marxism. (16,000)

3.3 THE COMMUNIST MOVEMENT, REVOLUTION, AND CIVIL WAR, 1921-49

Our original intention was to divide the material contained in this section into two parts, the first dealing with the history of China under the Kuomintang and the second with the history of the Communist movement itself. This, we soon found, was impossible to do because the only history of which the Chinese Communist historians take cognizance is that of the "struggles of the people," led by the Communist Party. No serious attention has been given to any aspect of the history of Kuomintang China apart from attacks on the reactionary character of the government and on the personal decadence of its leadership. Ch'en Po-ta's "biography" of Chiang Kai-shek (3.3.1) is an example of this genre. The currently orthodox periodization of China's "contemporary history" is as follows: 1921-27, "the first revolutionary civil war"; 1928-37, "the second revolutionary civil war"; 1937-45, the anti-Japanese patriotic war; and 1945-49, "the third revolutionary civil war." After a number of items (3.3.3-11) which deal specifically with the history of the Communist Party, the remaining titles in this section are grouped according to this periodization. See also 1.4.8, 1.4.15, 1.4.17, 1.4.19, and 1.4.21 among the surveys of modern and contemporary history discussed earlier in this volume, and 4.2.4-5.

In general, the current publications on the history of the Chinese Communist Party contain little or no new material. Obviously the archives of the Party are not open to historians, even to the Communist historians. Item 3.3.8 is a good example of Party history as it is now being written, with the emphasis principally on the public record of the Party and on the role of Mao Tse-tung as political leader and chief ideologist. The fragments of new information which appear, for example, in 3.3.9-11 are of marginal value and do not concern the principal problems in the Party's history. The next two items, on the People's Liberation Army and the Shanghai labor movement respectively, are also of little scholarly value.

The period of the "first revolutionary civil war" is covered by 3.3.14-22. These volumes are concerned in turn with the early history of the Chinese Communist Party, the peasant movement, and the early labor movement. There are apparently fewer materials available on the "second revolutionary civil war" (3.3.23-25). These are, of course, the critical years during which Mao Tse-tung gained the leadership of the Party and, profiting by the lessons of the defeat in 1927, harnessed it to the dynamic motor of peasant unrest, in contravention of the official line of the Party and of Moscow at the time. Perhaps this is still too touchy a subject for the mainland historians to treat. Items 3.3.26-32 are concerned with the war against Japan, and include two valuable works on the Communist-controlled border regions (3.3.29 and 3.3.30). The last group of titles (3.3.33-41) is concerned with the post-World War II struggle between the Communists and the Kuomintang, called the "third revolutionary civil war." In general, these volumes are propaganda rather than history.

3.3.1 Ch'en Po-ta 陳伯达 , Jen-min kung-ti Chiang Chieh-shih 人民公敵蔣介石 (Chiang Kai-shek, the people's public enemy; Peking: Hsin-hua shu-chü, 1954, first ed. 1948), 3 + 223 pp.

A widely circulated work in mainland China, this book is long on condemnation of Chiang Kai-shek and short on historical facts. Although the author's discussion of Chiang begins with the 1920's, there is no cohesive account of the progression of events in China which would provide a broader context or background. Instead, there are only references to isolated incidents which are introduced for the sole purpose of proving Chiang's undeviating villainy. Sources are cited infrequently, and sometimes imprecisely, in parentheses in the text. (75,000)

3.3.2 Meng Hsien-chang 孟憲章, Mei-kuo fu Chiang ch'in Hua tsui-hsing shih 美国扶蔣侵華罪行史 (A history of the criminal acts of the United States in supporting Chiang Kai-shek and carrying out aggression against China; Shanghai: Chung-hua shu-chü, 1951), 170 pp.

A work of unmitigated propaganda, with scaracely any scholarly pretentions, dealing with Sino-American relations briefly from the Open Door policy on, and in greater detail from the 1930's up to the Korean War. The presentation is calculated to incite the reader to hate America. (12,000)

3.3.3 Hu Ch'iao-mu 胡喬木, Chung-kuo kung-ch'an-tang ti san-shih nien 中国共産党的三十年 (Thirty years of the Communist Party of China; Peking: Jen-min ch'u-pan-she, 1951), 94 pp.

This official version of CCP history was written in 1951 in commemoration of the thirtieth anniversary of the Party's founding. It has gone through numerous printings in Chinese (the latest we have seen is the eleventh printing, June 1953, which brought the total number of copies in print to 1,480,000), and four printings to date in an English translation published by the Foreign Languages Press (the latest, May 1959, is described as a "revised English translation" made from a 1956 Chinese-language edition). The text glorifies Mao Tse-tung, who is described as an infallible theoretician and political and military strategist, whose leadership over thirty years brought the party and country to glory in spite of the obstruction of the "rightist" Ch'en Tu-hsiu 陳獨秀, the "leftist" Ch'ü Ch'iu-pai 瞿秋白, Chang Kuo-t'ao 張國濤, and others. Its organization of CCP history around the "three revolutionary civil wars" and the war against Japan is by now standard. (1,480,000)

3.3.4 Chi-nien Chung-kuo Kung-ch'an-tang ti san-shih chou-nien 紀念中国共產党的三十週年 (In commemoration of the thirtieth anniversary of the Chinese Communist Party; Hankow: Chung-nan jen-min ch'u-pan-she, 1951), 114 pp. (10,000)

3.3.5 Propaganda Department, South China Branch of the Central Committee of the CCP, ed., Chung-kuo Kung-ch'an-tang shih hsueh-hsi tzu-liao 中国共產党史学習資料 (Study material on the history of the Chinese Communist Party), No. 1 (Canton: Hua-nan jen-min ch'u-pan-she, 1952), 79 pp. (162,001)

3.3.6 Mou Ch'u-huang 繆楚黃, Chung-kuo Kung-ch'an-tang chien-yao li-shih, ch'u-kao 中国共产党簡要历史，初稿 (A brief history of the Chinese Communist Party, first draft; Peking: Hsueh-hsi tsa-chih she, 1956), 4 + 182 pp.

A textbook designed for a one-month course in lower party schools and cadre study groups. It covers the years 1921-49, and exaggerates the early role of Mao Tse-tung. It is, however, useful as a guide to what the CCP cadre is supposed to know. (200,200)

3.3.7 Huang Ho 黃河, Chung-kuo Kung-ch'an-tang san-shih-wu nien chien-shih 中国共产党三十五年簡史 (A brief history of thirty-five years of the Chinese Communist Party; Peking: T'ung-su tu-wu ch'u-pan-she, 1957), 96 pp.

A popular history of the CCP which narrates (with the appropriate clichés) the well-known highlights of the Party's history. Several interesting maps are included. (100,000)

3.3.8 Wang Shih 王实, et al., Chung-kuo Kung-ch'an-tang li-shih chien-pien 中国共产党历史簡編 (A concise history of the Chinese Communist Party; Shanghai: Shang-hai jen-min ch'u-pan-she, 1958), 6 + 302 pp.

A survey of CCP history from its inception to the establishment of the Chinese People's Republic in 1949 which stresses the role of Mao and implies his leadership of the party many years before this was an undisputed fact. The focus of the account is on the public role of the CCP, its declared policies, and their consequences, with very little attention to internal history or doctrinal problems. The text of this book was developed from teaching materials prepared for party cadres. No sources are cited except Mao Tse-tung's collected works; no bibliography is given. This text is widely used, as evidenced by the fact that the first printing, in September 1958, totaled 100,000 copies. (100,000)

3.3.9 Marxism-Leninism Night-school of the Wuhan Municipality, ed., Chung-kuo Kung-ch'an-tang tsai chung-nan ti-ch'ü ling-tao ko-ming tou-cheng ti li-shih tzu-liao 中国共产党在中南地区領導革命鬥争的历史資料 (Historical materials on the revolutionary struggles led by the Chinese Communist Party in Central and South China; Hankow: Chung-nan jen-min ch'u-pan-she, Vol. 1, 1951), 4 + 283 pp.

A collection of forty-one articles, most of which appeared in mainland newspapers in 1951. They are arranged under the following headings: (1) general accounts of revolutionary struggles in the Central and Southern areas; (2) revolutionary struggles in the periods of the formation of the CCP and the first "revolutionary civil war"; and (3) revolutionary struggles during the period of the "second revolutionary civil war." A number of the articles are reminiscences by participants; and much of the material is mere propaganda. (10,000)

3.3.10 Chung-kuo Kung-ch'an-tang ling-tao Hu-nan jen-min ying-yung fen-tou ti san-shih nien 中国共产党領導湖南人民英勇奮鬥的三十年 (Thirty years of the Hunan people's courageous struggles led by the Chinese Communist Party; Changsha: Hu-nan t'ung-su tu-wu ch'u-pan-she, 1951), 72 pp. (30,000)

3.3.11 Hua Ying-shen 華應申, Chung-kuo Kung-ch'an-tang lieh-shih chuan 中国共产党烈士伝 (Biographies of Chinese Communist martyrs), expanded ed. (Peking: Ch'ing-nien ch'u-pan-she, 1951), 6 + 260 pp. (15,000)

3.3.12 Huang T'ao 黄濤, Chung-kuo jen-min chieh-fang-chün ti san-shih nien 中国人民解放軍的三十年 (Thirty years of the Chinese People's Liberation Army: Peking: Jen-min ch'u-pan-she, 1958), 60 pp.

A popular pamphlet written in commemoration of the thirtieth anniversary of the establishment of the Chinese Communist army. Its four chapters consider in turn the period of the Nationalist Revolution and the Kiangsi Soviet, the war against Japan, the KMT-CCP civil war, and (more briefly) the army since 1949. The protagonist of this narrative is as much Mao Tse-tung as it is the People's Liberation Army. (74,500)

3.3.13 Cultural and Educational Department of the Shanghai Federation of Labor, ed., San-shih nien lai ti Shang-hai kung-yün 三十年来的上海工運 (The Shanghai labor movement over the past thirty years; Shanghai: Lao-tung ch'u-pan-she, 1951), 28 pp. (8,000)

3.3.14 Wang K'o-feng 王可风, Wu-ssu yun-tung yü Chung-kuo Kung-ch'an-tang ti tan-sheng 五四运动与中国共产党的誕生 (The May Fourth Movement and the birth of the Chinese Communist Party; Nanking: Chiang-su jen-min ch'u-pan-she, 1958), 43 pp.

The May Fourth Movement and the organization of the CCP are coupled in this pamphlet in order to show the latter both as the real culmination of the former, and as a token that the Chinese revolution had entered onto a new and higher stage. (15,000)

3.3.15 Wei Hung-yun 魏宏运, Pa-i ch'i-i 八一起义 (The uprising of August 1 [1927]; Wuhan: Hu-pei jen-min ch'u-pan-she, 1957), 32 pp.

A pamphlet dealing with three important events in the history of the Chinese Communist movement: (1) the Nanchang uprising of August 1, 1927 led by Ho Lung 賀龍 and Yeh T'ing 葉挺; (2) the advance of the Communist forces into Ch'ao-chou and Swatow, Kwangtung, in September of that year; and (3) the union of the forces of Mao Tse-tung and Chu Teh at Ching-kang-shan 井岡山, Hunan, and the formation of the Red Fourth Army in May 1928. A fair number of sources are cited in footnotes. (20,000)

3.3.16 Ti-i-tz'u kuo-nei ko-ming chan-cheng shih-ch'i ti chi chien shih-shih 第一次国内战争時期的几件史实

(Some historical facts concerning the first revolutionary civil war; Shanghai: Hsin chih-shih ch'u-pan-she, 1956), 68 pp. (25,000)

3.3.17 Yeh Hu-sheng 葉蠖生 and Lo Yang-sheng 羅仰生 , Ti-i-tz'u kuo-nei ko-ming chan-cheng chien shih 第一次国内革命战争簡史 (A brief history of the first revolutionary civil war; Shanghai: Shang-hai jen-min ch'u-pan-she, 1957), 2 + 132 pp., 8 plates. (26,000)

3.3.18 Ti-i-tz'u kuo-nei ko-ming chan-cheng shih-ch'i ti nung-min yun-tung 第一次国内革命战争時期的农民运动 (The peasant movements in the period of the first revolutionary civil war; Peking: Jen-min ch'u-pan-she, 1953), 5 + 439 pp.

This collection of source materials on Chinese peasant conditions and movements in 1926-27 is divided into two unequal parts. Part I (pp. 3-32) contains four selections from KMT periodicals devoted to the "peasant problem," one of which deals with the training of peasant-movement cadres. Part II (pp. 35-439) contains one section each on Kwangtung, Hunan, and "other provinces." (In this last, materials are quite meager on all but Hupei and Kiangsi.) The bulk of the collection concerns Kwangtung and Hunan, where the peasant movement was most active. The principal sources are KMT and CCP periodicals. Proclamations and other documents of the rural peasant associations are included, and other selections range from a presumably firsthand account of the peasant movement in Hai-feng, Kwangtung, in 1921-23 to a research article on the peasant movement in Hunan published in the journal Hsueh-hsi 學習 in 1951. (25,000)

3.3.19 Chung I-mou 鍾貽謀 , Hai-Lu-feng nung-min yun-tung 海陸豐農民運動 (The peasant movement in Hai-feng hsien and Lu-feng hsien [Kwangtung province, ca. 1922-25]; Canton: Kuang tung jen-min ch'u-pan-she, 1957), 128 pp., 6 plates. (6,120)

3.3.20 Teng Chung-hsia 鄧中夏 , Chung-kuo chih-kung yun-tung chien shih, 1919-1926 中国職工運動簡史 (A concise history of the Chinese labor movement, 1919-1926; Peking: Jen-min ch'u-pan-she, 1953), 8 + 259 pp.

This book was written between 1928 and 1930 in Moscow, where the author had gone as a delegate to the Sixth Congress of the Communist International. It was published in Moscow in 1930 and a first Chinese edition appeared in 1943. As Teng (1894-1933) was a principal Communist labor leader in the 1920's, his account is on the whole authoritative, but for this very reason it tends to overstate the ideological consciousness, cohesiveness, and organizational effectiveness of the strikes and other activities he describes. On balance, the book contains useful material on the major strikes in China and Hong Kong between 1922 and 1926, on the organization of the First and Second Labor Congresses; and on the formation of the All China Labor Federation. A bgiographical sketch of Teng Chung-hsia appears as an appendix. (40,000)

3.3.21 Liu Li-k'ai 刘立凱 and Wang Chen 王真 , I-chiu-i-chiu chih i-chiu-erh-ch'i nien ti Chung-kuo kung-jen yun-tung 一九一九至一九二七年的中国工人运动 (The Chinese labor movement between 1919 and 1927; Peking: Kung-jen ch'u-pan-she, 1953), 61 pp.

This small book combines two articles previously published in Hsueh-hsi magazine which give an official account in capsule form of the major labor strikes between the May Fourth Movement and the Northern Expedition, emphasizing the role of the CCP. (50,084)

3.3.22 Ti-i-tz'u kuo-nei ko-ming chan-cheng shih-ch'i ti kung-jen yun-tung 第一次国内革命战争時期的工人運動 (The Chinese labor movement in the period of the first revolutionary civil war; Peking: Jen-min ch'u-pan-she, 1954), 5 + 556 pp.

A collection of source materials consisting largely of articles from contemporary periodicals (among them, organs of the CCP, KMT, and the trade union federation) covering the period 1924-31 and illustrating the mobilization of the labor movement in the "anti-imperialist, anti-warlord" cause. The materials concentrate on strikes against foreign-owned enterprises as well as on such "incidents" as the May thirtieth shootings and the Shakee Massacre of 1925. Among other things, there are articles on the KMT-CCP alliance, the Canton-Hongkong strike of 1925-26, the role of the workers in the Northern Expedition, the 1927 Shanghai coup d'état, and the subsequent suppression of the labor movement in Canton and elsewhere. The nature of the materials varies from firsthand accounts by participants to official pronouncements of labor congresses and the CCP. (23,000)

3.3.23 Ch'en Po-ta 陳伯達, Kuan-yü shih nien nei-chan 关于十年内战 (Concerning ten years of civil war [ca. 1927-37]; Peking: Jen-min ch'u-pan-she, 1953), 69 pp. (10,000)

3.3.24 Wang Po-yen 汪伯岩 , Ti-erh-tz'u kuo-nei ko-ming chan-cheng shih-ch'i ti nung-ts'un ko-ming ken-chü-ti 第二次国内革命战争時期的農村革命根據地 (The rural revolutionary base during the "second revolutionary civil war"); Shanghai: Hsin-chih-shih ch'u-pan-she, 1956 [first publ. 1955]), 138 pp.

A study of the Kiangsi Soviet period of CCP history, based largely on the collected writings of Mao Tse-tung. The narrative covers the years 1928-43 and, while it gives some attention to other ephemeral Communist guerrilla bases, concentrates on the relatively long-lived Kiangsi-Hunan border area. Although useful for details on the movements of the CCP forces, the book is otherwise very thin fare because, among other reasons, the author has obviously not had access to such records as have survived from this period. (48,000)

3.3.25 "Historical Bi-monthly" Society, ed., Ti-erh-tz'u kuo-nei ko-ming chan-cheng shih-ch'i shih-shih lun ts'ung 第二次国内革命战争時期史事論丛 (Collected essays on historical facts during the "second revolutionary civil war"; Peking: San-lien shu-tien, 1956), 118 pp. (15,000)

3.3.26 I-erh-chiu yun-tung 一二九运动 (The December ninth [1935] movement; Peking: Jen-min ch'u-pan-she, 1954), 2 + 158 pp.

Sixteen articles, mostly from periodicals of the year 1936 (with some of a much later date) are reprinted in this collection dealing with the anti-Japanese student demonstrations of December 1935 in Peking, Tientsin, and other cities. The texts of several manifestoes by student and other groups on this occasion are also included. The publisher's note claims that the December 9 movement was led by the CCP, as was the Peiping Student Union which staged the demonstrations. That the CCP leadership in the movement is not brought out in the material reprinted here is attributed to the fact that the periodicals represented were all published in KMT areas. (25,000)

3.3.27 K'ang-Jih chan-cheng shih-ch'i ti Chung-kuo jen-min chieh-fang-chün 抗日战争时期的中国人民解放軍 (The Chinese People's Liberation Army during the period of the war against Japan; Peking: Jen-min ch'u-pan-she, 1953), 2 + 231 pp., illus.

This is a slightly revised version of a book called K'ang-chan pa-nien lai ti Pa-lu-chün ho Hsin-ssu-chün 抗战八年来的八路軍和新四軍 (The Eighth Route Army and New Fourth Army in eight years of war of resistance), published in 1945 by the Propaganda Department of the Eighth Route Army. It contains an account of the campaigns of these two armies and of guerilla warfare in North and South China between 1937 and 1944. A short appendix discusses in general terms the militia in Communist-held areas during this period. (30,000)

3.3.28 K'ang-Jih chan-cheng shih-ch'i ti chieh-fang-ch'ü kai-k'uang 抗日战争時期解放区概况 (Conditions in the "liberated areas" during the Sino-Japanese War; Peking: Jen-min ch'u-pan-she, 1953), 3 + 132 pp., 8 maps.

A description of the Chinese Communist-held areas between 1937 and 1945. Such topics as the establishment and administration of border areas, guerrilla warfare, and local economic, social, and cultural life are described for each of the four main "liberated areas": the Shensi-Kansu-Ninghsia Border Area, and the areas behind enemy lines in North China, Central China, and South China. The text at times is quite detailed and contains many statistics, but the sources used are not indicated. The maps are informative. (20,000)

3.3.29 Third Office, Institute of Historical Research, Chinese Academy of Sciences, comp., Shan-Kan-Ning pien-ch'ü ts'an-i-hui wen-hsien hui-chi 陕甘宁边区参议会文献汇辑 (A documentary collection on the Shensi-Kansu-Ninghsia Border Region People's Council; Peking: K'o-hsueh ch'u-pan-she, 1958), vi + 379 pp., 3 plates.

This volume contains records of four meetings held between 1939 and 1946 by the People's Council of the Shensi-Kansu-Ninghsia Border Region, headquarters of the CCP during China's war with Japan. The records reproduced here were originally published separately by the People's Council or by the Border Region government. They consist of the agenda, reports, proposals, resolutions and programs in their entirety as passed by each of the Council sessions, and are therefore an important group of source materials on CCP policies and administration in the Yenan period. (4,689)

3.3.30 Ch'i Wu 齐武, I-ko ko-ming ken-chü-ti ti ch'eng-chang: K'ang-Jih chan-cheng ho chieh-fang chan-cheng shih-ch'i ti Ch'in Chi Lu Yü pien-ch'ü kai-k'uang 一个革命根据地的成長：抗日战争和解放战争时期的晋冀鲁豫边区概况 (The establishment and growth of a revolutionary base: A general account of the Shansi-Hopei-Shantung-Honan border region during the anti-Japanese war and the war of liberation; Peking: Jen-min ch'u-pan-she, 1958), 3 + 323 pp., 6 plates, 1 map.

The history of the Shansi-Hopei-Shantung-Honan border region from 1937 through 1948 begins with an introductory section on the geography of the region and its administrative organization (including a list of the administrative areas, special districts, etc., and the places under their jurisdiction in 1946). The succeeding chapters give a chronological account of the establishment and growth of the "revolutionary base" (1937-40), its consolidation and persistence in the face of KMT attacks (1941-42), and its role in the war against Japan and during the subsequent civil war. The narrative of conditions in the border region in these years is relatively detailed, and provides considerable, though partisan, information about Communist economic, political, and military policy. While a number of notes in the text indicate the sources used, many parts of the text are undocumented. Explanatory notes at the end of the volume help identify events referred to in the text. The "Revised, provisional regulation governing land use in the border region" (1942) is reproduced on pages 310-313. (18,500)

3.3.31 Ch'en Ts'ung-i 陳從一, Wan-nan shih-pien ch'ien-hou 皖南事变前后 (Before and after the Southern Anhwei Incident; Shanghai: Shang-hai jen-min ch'u-pan-she, 1950), 74 pp., 3 plates.

On the ambush of contingents of the Communist New Fourth Army by KMT troops on January 4, 1941, in which ca. 9,000 CCP troops were wiped out and their leaders captured or killed. Events prior to this incident, and its consequences, are also detailed in this brief volume. (93,000)

3.3.32 Feng Pai-chü 馮白駒, et al., Kuang-tung jen-min k'ang-Jih yu-chi chan-cheng hui-i 廣東人民抗日游击战争回忆

(Reminiscences on the Kwangtung people's anti-Japanese guerrilla warfare; Canton: Hua-nan jen-min ch'u-pan-she, 1951), 30 pp. (10,000)

3.3.33 Ti-san-tz'u kuo-nei ko-ming chan-cheng kai-k'uang 第三次国内革命战争概况 (A general account of the third revolutionary civil war; Peking: Jen-min ch'u-pan-she, 1954), 2 + 243 pp., 4 maps. (30,000)

3.3.34 Ch'üan Wei-t'ien 全慰天, Ts'ung chiu Chung-kuo tao hsin Chung-kuo -- ti-san-tz'u kuo-nei ko-ming chan-cheng shih-ch'i ching-chi-shih lueh 從旧中国到新中国--第三次国内革命战争时期經济史略 (From old China to new China--a brief economic history of the period of the third revolutionary civil war [1945-49]; Shanghai: Hsin chih-shih ch'u-pan-she, 1957), 4 + 470 pp.

> The publisher's "summary of the contents" which appears on the back cover is indicative of the nature of this book. It says: "This volume may be read as a reference on Chinese revolutionary history and Chinese economic history; it may also serve as political reading matter for the general public." To accomplish the latter purpose, the collapse of the KMT economy and the development of Communist economy in the "liberated areas" between 1945 and 1949 are presented as the inevitable result of Marxian economic law. However, there is a considerable amount of data on the economy of the "liberated areas" tucked away within the tendentious narrative. (8,000)

3.3.35 Yeh Hu-sheng 葉蠖生, Chung-kuo jen-min chieh-fang chan-cheng chung ti li-shih ku-shih 中国人民解放战争中的历史故事 (Historical tales of the Chinese people's war of liberation;

Shanghai: Hua-tung jen-min ch'u-pan-she, 1954), 61 pp. (165,000)

3.3.36 Mu Hsin 穆欣, Nan hsien hsün-hui 南線巡迴 (Inspections of the southern front; Peking: San-lien shu-tien, 1951), 6 + 211 pp., illus., map.

Originally written and published as a newspaper serial, this is a story of the southward advances of the Fourth Regiment, Second Field Army, Chinese People's Liberation Army, between 1948 and 1950. The author appears to be a reporter who traveled with this regiment and observed it in action, but it is difficult to tell from the narrative how much of the account is actually firsthand. The book is divided into three main parts, within which there are numerous subdivisions. Part I describes the crossing of the Yangtze river; Part II treats of events as the regiment reached the original "Soviet areas" in Kiangsi, Hupei, and Hunan; and Part III consists of accounts of the victorious advance southwestward into Kwangtung, Kwangsi, Kweichow, and Yunnan. The dominant theme is the greatness of the Red Army and the people's love for it. The narrative is accompanied by photographs, and there is an excellent map. (40,000)

3.3.37 Hu En-tse 胡恩沢, Hui-i ti-san-tz'u kuo-nei ko-ming chan-cheng shih-ch'i ti Shang-hai hsueh-sheng yun-tung 回忆第三次国内革命战争时期的上海学生运动 (Reminiscences of the Shanghai student movement during the "third revolutionary civil war"; Shanghai: Shang-hai jen-min ch'u-pan-she, 1958), 60 pp.

The author has drawn on his own experiences, newspaper accounts, and "illegal" student publications in order to compile these highly colored accounts of student demonstrations in Shanghai between 1945 and 1949. Thirteen incidents are described in varying degrees of detail. (10,000)

3.3.38 Liao Kai-lung 廖蓋隆, Chung-kuo jen-min chieh-fang chan-cheng ho hsin Chung-kuo wu-nien chien-shih 中国人民解放战争和新中国五年簡史 (A brief history of the Chinese People's war of liberation and five years of the New China; Peking: Jen-min chiao-yü ch'u-pan-she, 1955), 4 + 176 pp.

The 1955 edition of a textbook for middle school upper grades originally published in 1951. Chapters 1-4 cover the period 1945-49 (pp. 1-95), and chapters 5-7 deal respectively with the establishment of "New China," rehabilitation of the economy, and the inauguration of the first Five Year Plan. "Questions for review" appear at the end of each chapter. (30,000)

3.3.39 Liao Kai-lung 廖蓋隆, Chung-kuo jen-min chieh-fang chan-cheng chien-shih 中国人民解放战争簡史 (A brief history of the Chinese people's war of liberation; Shanghai: Hai-yen shu-tien, 1952, originally pub. 1951), 4 + 134 pp.

An earlier version of the preceding item, covering the years 1945-49. (110,000)

3.3.40 Liao Kai-lung 廖蓋隆, Hsin Chung-kuo shih tsen-yang tan-sheng ti 新中國是怎樣誕生的 (The birth of New China), expanded ed. (Shanghai: Hai-yen shu-tien, 1950), 7 + 271 pp. (45,000)

3.3.41 Chung-hua jen-min kung-ho-kuo k'ai-kuo wen-hsien 中華人民共和国開国文献 (Historical records of the beginning of the Chinese People's Republic; Hong Kong: Hsin min-chu ch'u-pan-she, 1949), 10 + 311 pp.

This volume is a record of the major events and speeches at the First Session of the People's Political Consultative Conference, September 21-30, 1949. All the items in Part I, including Chairman Mao's opening address, reports by other CCP leaders, lists of representatives to the conference, and day-by-day summaries of the sessions, are Hsinhua News Agency dispatches. Part II reproduces a series of speeches by representatives of various parties, districts, armies, organizations, and "special invitees." Part III contains resolutions passed by the conference. Part IV, the appendix, contains four editorials from official Communist Chinese and Soviet newspapers. (5,000)

4. ECONOMIC HISTORY

The works on economic history in this section are arranged by subject rather than chronologically. They contain materials relevant to many of the periods and topics discussed in earlier sections of this bibliography. On the other hand, given the Marxist orientation of nearly all of the volumes we have surveyed, and the consequent emphasis on economic "causes," materials on the history of the Chinese economy will also be found in the "political histories" included in Sections 1, 2, and 3. The Marxist emphasis on the economic "infrastructure" of society also implies that the study of economic history is highly valued and greatly encouraged. In mid-1958, for example, it was reported that thirty-nine major projects to collect and compile source materials on the modern economic history of China were under way. A large number of research institutions had each assumed the responsibility for one or more topics, and 1961 was set as the target date for completing these compilations (see Ching-chi yen-chiu 經濟研究 [Economic research], pp. 89-90 [May 1958]). It may, therefore, be assumed that the titles included in this part are only the first wave of a large flood of documentary collections and monographic studies of modern economic history.

It is clear that in the future, Western scholars of this subject will be increasingly dependent on the output of historical materials from mainland China. At present these fall roughly into three categories of which there are examples in the individual titles discussed below. At one extreme are the doctrinaire and schematic works whose principal purpose is to fit China's economic history into the Marxist normative framework of development. At the other extreme are such publications as the price indices in sec. 4.5 which provide invaluable and undoctored data (except perhaps for a tendentious preface). In between fall the increasingly numerous source collections, arranged usually to support the Marxist argument but containing valuable data often not obtainable elsewhere. Many fields of research are being "opened" and the foundation laid--if the cruder Marxist analysis can be set aside--for writing the economic history of modern China at a higher level of theoretical sophistication and with a more comprehensive control of the empirical data than was the case in the past.

4.1 SURVEYS AND MATERIALS

Economic history in Communist China is directed toward two ends: first, to demonstrate that the Chinese economy was evolving through the same stages which Marx asserted it had passed through in the West, and second, to place the blame for China's failure to develop into a full-fledged capitalist nation on the aggression of foreign imperialism. With the exception of Li Chien-nung's traditional survey of Chinese economic history (4.1.4-6) the studies and collections of source materials discussed in this section do not depart from these guidelines. The documentation employed, as for example in the first three items, is often impressive, but the analysis is thin. As in the foregoing sections and in those below, the collection and publication of source materials is probably the most valuable service being performed by the mainland historians. Item 4.1.8, in particular, is an extremely useful research tool for China's modern economic history. Yen Chung-p'ing's statistical handbook (4.1.9) is also deserving of close attention. The introductory materials, which were probably written by Yen himself, are the product of a relatively sophisticated writer. Yet if this collection of statistics is really the best of what is available for China's economic history in the past century, economic historians face a difficult prospect. Survey treatments of economic history are also included in the works discussed in secs. 1.1 and 1.4 above. See also 3.3.34.

4.1.1 K'ung Ching-wei 孔經緯 , *Chung-kuo ching-chi shih shang ti chi-ko wen-t'i* 中国經济史上的几个問題 (Some problems of Chinese economic history; Shanghai: Shang-hai jen-min ch'u-pan-she, 1956), 64 pp.

Four rather thin papers: "The evolution of the T'ang land and tax systems" (pp. 1-11); "The growth of the economic power of the bourgeoisie in Hangchow in the Southern Sung" (pp. 12-22); "The development of national capitalist modern industry and its level prior to 1937" (pp. 23-47), based almost entirely on Kung Chün's 龔駿 now outdates survey: *Chung-kuo hsin kung-yeh fa-chan shih ta-kang* 中国新工業發展史大綱 (Outline history of the development of Chinese modern industry; Shanghai, 1933); and "The question of the formation of a national market" (pp. 48-64)--a mere shadow of Lenin's *The Development of Capitalism in Russia* on which it is modeled. (8,000)

4.1.2 Shang Yueh 尚鉞 , ed., *Chung-kuo feng-chien ching-chi kuang-hsi ti jo-kan wen-t'i* 中国封建經济关系的若干問題 (Some problems of feudal economic relations in China; Peking: San-lien shu-tien, 1958), 6 + 345 pp.

Nine articles prepared at the Chinese People's University in Peking as part of the history department's 1955-56 "plan," with an introduction by Professor Shang Yueh who is obviously the ideological inspiration for the pieces. The time span covered is from the end of the Han to the early Ch'ing, and the topics include Ts'ao Ts'ao's 曹操 military agricultural colonies; relations of production in the Eastern Chin and the Southern Dynasties; the land system of the Northern Wei; T'ang landed estates, handicraft industry, and labor service; Yuan handicraft industry; Li Tzu-ch'eng's 李自成 land equalization policy; and early Ch'ing urban uprisings. Following in general the position that Shang Yueh has taken in the incessant periodization discussions (see sec. 4.3), these essays in sum hold that "feudalism" in China began at the end of the later Han, reached its peak in the T'ang, and declined from the Ming onwards in the face of growing capitalist burgeons (tzu-pen chu-i meng-ya 資本主義萌芽). Although sometimes crude in their argument, the essays are well documented and at least indicate how far it is possible to go in placing the evolution of Chinese society within a framework taken over wholesale from the Western European experience. (3,700)

4.1.3 Han Kuo-p'an 韓國磐, Pei-ch'ao ching-chi shih-tan 北朝經济試探 (An inquiry into the economy of the Northern Dynasties; Shanghai: Shang-hai jen-min ch'u-pan-she, 1958), 226 pp.

The five topical chapters in this volume are based primarily on the standard histories (especially the Wei-shu 魏書) of the Northern Dynasties (336-588). They treat the following subjects: (1) the economic changes attending the establishment of the T'o-pa state, which the author sees as a transition to Chinese-type "feudalism" from an antecedent "tribal" stage; (2) the property and tax systems of the Wei dynasty; (3) the origins of the chün-t'ien 均田 ("equal land allotment") system, supplementing the author's earlier study Sui T'ang ti chün-t'ien chih-tu 隋唐的均田制度 (The equal land allotment system of the Sui and T'ang dynasties), not seen by us; (4) industry and commerce, including a consideration of the degree of monetization of the economy; and (5) the economy of the Buddhist monasteries in the Northern Dynasties. The descriptive data extracted from the sources are useful, but the analysis is a very elementary brand of Marxism. (3,000)

4.1.4 Li Chien-nung 李劍農, Hsien-Ch'in liang-Han ching-chi shih-kao 先秦兩漢經濟史稿 (Draft economic history of China through the Han dynasty; Peking: San-lien shu-tien, 1957), 291 pp.

This is a new edition, with minor corrections, of Professor Li's pre-1949 lectures at Wuhan University on the economic history of traditional China. A limited edition, for student use, was published in 1948: Chung-kuo ching-chi shih chiang-kao 中國經濟史講稿 (Lectures on the economic history of China), Vol. 1, distributed by the president, National Wuhan University, Wuchang. The author's preface to the present volume states that although his ideas about the society of ancient China have since changed, he is nearly blind and too feeble to undertake a major revision. Therefore, unlike most post-1949 writing on this subject, this book is not preoccupied with the problem of periodizing ancient history in order to determine when the period of slavery ended and that of feudalism began. Its eighteen chapters give a careful survey, based on the primary sources, of the key topics in the economic history of the period from the Shang through the end of the Later Han. For example, Part 3, "The Two Han Dynasties" (pp. 141-291), covers the following subjects: the improvement of agriculture, industry, money, commerce, the land system, taxation, and economic thought and policies. On many of the issues that trouble the Marxist historians, Professor Li is neutral and academic. Note that throughout the term feng-chien 封建 (used for "feudalism" today though itself an ancient term) is used in a more technical sense than is now the case in mainland historiography in general. (5,500)

4.1.5 Li Chien-nung 李劍農, Wei Chin Nan-pei-ch'ao Sui T'ang ching-chi shih kao 魏晋南北朝隋唐經濟史稿

(Draft economic history of the period A.D. 220-906; Peking: San-lien shu-tien, 1959), 3 + 302 pp.

A continuation of the previous item. One hundred-odd copies of the first draft were mimeographed in 1943 for student use. After 1945, Professor Li made some revisions in the first five chapters, but chapters 7-13 are basically unchanged from the draft. Chapter 6 was written by P'eng Yü-hsin 彭雨新 who assisted with the editing of the present version. Like the previous item, the text is detailed, replete with long quotations from the sources, and free of Marxist preoccupation. "Economic history," with reference to these three volumes by Professor Li, is to be understood as "the history of certain economic institutions"; there is little economic analysis. (2,000)

4.1.6 Li Chien-nung 李劍農, <u>Sung Yuan Ming ching-chi shih kao</u> 宋元明經濟史稿 (Draft economic history of the period 960-1644 A.D.; Peking: San-lien shu-tien, 1957), 2 + 297 pp.

Like the previous two items, an expansion of Professor Li's lecture notes at Wuhan University which was originally issued in mimeographed form for student use. Again the author's preface admits that from a theoretical point of view these chapters are inadequate, i.e., that they are not Marxian. While Chapter 3 does refer to the development of handicraft industry in the Sung, Yuan, and Ming dynasties, the point of departure is not the now-accepted <u>tzu-pen</u> <u>chu-i</u> <u>meng-ya</u> ("capitalist burgeons") controversy. (15,000)

4.1.7 Department of Political Economy, Hupei University, ed., <u>Chung-kuo chin-tai kuo-min ching-chi shih chiang-i</u> 中国近代国民經济史講义 (Lectures on the modern national economic history of China;

Peking: Kao-teng chiao-yü ch'u-pan-she, 1958), 12 + 546 pp.

The ten chapters of this textbook are written by four different persons, one of whom is a professor at Wuhan University. As a "preliminary investigation" into the course of economic development and chnage in China in the last 100 years, this book, say the editors, will answer such questions as these: "What were the stages in the transformation of China from a feudal society to a semi-colonial and semi-feudal society?" "How was the semi-colonial, semi-feudal nature of the economy intensified and eventually destroyed?" "What are the characteristics of each of the stages of economic development and change?" "What about the class struggles that accompanied economic change?" These questions, plus the repeated references in the editors' preface to Mao Tse-tung's dicta on the Chinese economy and the impact of imperialist aggression, are a fair indication of the conclusions supplied. However, this is probably the most detailed consecutive treatment of its subject now available, and since it is reasonably well-documented, it should be of considerable value for a preliminary reconnaissance of the economic history of China in the nineteenth and twentieth centuries. (5,000)

4.1.8 Department of History, Nankai University, comp., Ch'ing shih lu ching-chi tzu-liao chi-yao 清實錄經濟資料輯要 (A compendium of economic materials in the Ch'ing shih lu; Peking: Chung-hua shu-chü, 1959), 1009 pp.

The materials relating to the development of the Chinese economy that appear in the Ch'ing shih lu and the Hsüan-t'ung cheng chi 宣統正紀 (covering the period 1583-1911) have been extracted by the editors and brought together in this volume. They are arranged chronologically under the following major headings which in turn are divided into smaller categories: (1) general, (2) agriculture, (3) animal husbandry, (4) handicrafts, (5) modern industry, (6) transport and communications, (7) commerce and usury, (8) foreign trade (to which is appended "the invasion of foreign capital"), (9) public finance, (10) taxation, (11) salt administration, and (12) grain transport. Except for the addition of punctuation and substitution of the "man" radical for the pejorative "animal" radical in words referring to non-Chinese minority peoples, the passages are given as they appear in the original, with the date, chüan, and page number at the end of each item. However, the frequent use of ellipses to indicate the omission of materials of a non-economic nature makes it necessary to refer to the original for the context in which many of the extracts occur. This volume, in fact, is most useful as a kind of index to material on the economy in the Ch'ing Shih lu. (1,100)

4.1.9 Yen Chung-p'ing 嚴中平 et al., comps., Chung-kuo chin-tai ching-chi shih t'ung-chi tzu-liao hsüan-chi 中国近代經濟史統計資料選輯 (A collection of statistical materials on the modern economic history of China; Peking: K'o-hsueh ch'u-pan-she, 1955), 15 + 374 pp.

A volume much cited by CCP writers on China's modern history. It contains 250 statistical tables covering the period 1840-1948 (the foreign trade statistics, however, go back to 1760), and the following topics: Anglo-Chinese trade prior to the Opium War (31 tables); treaty ports and concessions (3 tables); foreign trade (31 tables); industry (47 tables); railroads (30 tables); shipping (23 tables); and agriculture (85 tables).

Each of these is divided into subtopics (e.g., under agriculture: rural class structure, land ownership, land utilization, rents, tenancy, and extra-economic compulsion, changes in the conditions of tenancy, the commercialization of the economy, rural usury, and agricultural output). Each section is prefaced with a brief discussion of the significance of the statistics that follow. For the most part the data are taken from published materials in Chinese (those concerning pre-1840 Anglo-Chinese trade are largely from Earl Pritchard, The Crucial Years of Anglo-Chinese Relations, 1750-1800 or from H. B. Morse, The Chronicles of the East India Company Trading to China), and the sources of each table are carefully indicated. As might be expected, the data are fuller for more recent years than for the nineteenth century. An appendix (pp. 362-374) prints annual population figures by province for the years 1786-1898 (with gaps) from Hu-pu reports in the Ch'ing archives. Close examination of the tables shows that many are trivial in content, that the "index number problem" has not been faced squarely by the compilers, that the data are often too fragmentary to support some of the sweeping statements about the "semi-feudal, semi-colonial" character of the Chinese economy in the introductory sections (the tone of which is, however, surprisingly moderate), and that, in general, adequate and reliable statistics about the Chinese economy prior to 1949 are lacking. (4,390)

4.2 AGRICULTURE, LAND TENURE, AND TAXATION

Perhaps because they are so intimately a part of "feudal" society (which to the mainland historians is coincident with Imperial China), relatively less attention seems to have been given to agricultural history, land tenure, and land taxation than to the problems of incipient capitalism and the beginnings of modern industry. As we have seen (sec. 1.2 above) there has been a reluctance to venture forth in the uncharted waters of the period between the Han Dynasty and the late Ming Dynasty, perhaps because the Marxist classics have had little to say about the anatomy of feudalism and it is potentially dangerous to the Communist historian to be faced with ambiguous guidelines. Items 4.2.1 and 4.2.2 are interesting and valuable exceptions to this general lacuna. See also the general surveys and interpretations in sec. 1.1 above. Items 4.2.3-5 provide an enormous collection of source materials on China's modern agricultural history from 1840 to 1937. While the volumes are tendentiously arranged and lack the introductory essays that are found in the volumes on modern industry, they are compiled from an immense variety of often inaccessible sources and can be the starting place for a much needed study of modern Chinese agriculture. Despite the Marxist framework--or perhaps because of it--they are still very much in the scissors-and-paste tradition that has an honored place in Chinese historiography. The shortcomings in these volumes (and in the parallel collections for industry and handicraft) stem in part from the nature of the materials on which the compilers have relied. It is, nevertheless, this dependence on traditional sources and their citation in profusion (and with footnotes) that is also probably the chief value to Western scholars of the recent writing from mainland China on early modern economic history.

4.2.1 Ho Ch'ang-ch'ün 賀昌群, *Han T'ang chien feng-chien ti kuo-yu t'u-ti chih yü chün-t'ien chih* 漢唐間封建的国有土地制与均田制 (Feudal state land ownership and land equalization in the period from the Han to the T'ang; Shanghai: Shang-hai jen-min ch'u-pan-she, 1958), 130 pp.

A detailed study of the background of the T'ang land

equalization system, stressing its relation to the "feudal" institution of state-owned land (kung-t'ien 公田). The period covered is roughly the mid-Han to the mid-T'ang. The author quotes profusely from the chief Chinese sources and--something unusual in historical writing from the Chinese People's Republic--makes rather frequent reference to Japanese work on this subject, especially to that of Professor Niida Noboru 仁井田陞. One of his principal conclusions is that land equalization schemes were put into practice only in regions that were under direct imperial control, i.e. in the environs of the capital, but also in the country at large on those few occasions in the Northern Wei, the Sui, and the T'ang when imperial power was at its peak. The class struggle goes on throughout the volume, but not obtrusively. There are quite a number of interesting hypotheses put forward that might well be further investigated. (3,000)

4.2.2 Liang Fang-chung 梁方仲, Ming-tai liang-chang chih-tu 明代糧長制度 (The "land-tax collector" system of the Ming dynasty; Shanghai: Shang-hai jen-min ch'u-pan-she, 1957), 148 pp.

First published in a historical journal in 1935, the original manuscript of this scholarly book has gone through four revisions (for the third revision, see Ming Ch'ing shih lun ts'ung, item 2.1.2). The present version, enlarged to some 90,000 characters, discusses: (1) the origin and raison d'être of the collector of land tax, (2) the functions and special privileges of that office, (3) the evolution of the system, and (4) class differentiation among the collectors of land tax and their exploitation of the people. This is a detailed, systematic, and analytic study based on the Ming histories, local gazetteers, and much other literature of the Ming period. There is not a single mention of Marx, Engels, Lenin, Stalin or Mao, in the text or notes. Among the author's previous works is the authoritative I-t'iao-pien fa 一條鞭法 (1936), available in English translation as The Single-whip Method of Taxation in China (Harvard University Press, 1956). (10,000)

4.2.3 Li Wen-chih 李文治, comp., Chung-kuo chin-tai nung-yeh shih tzu-liao, ti-i chi, 1840-1911 中国近代农业史資料第一輯 (Source materials on China's modern agricultural history, first collection, 1840-1911; Peking: San-lien shu-tien, 1957), 21 + 1023 pp.

Excerpts from Ch'ing documentary collections, gazetters, agricultural works, private writings of all kinds, modern studies, newspapers and magazines, and some archival materials (all listed in a bibliography of ca. 600 titles [pp. 1003-23]), illustrative of Chinese agriculture at the end of the Manchu dynasty. As in the other collections in this series, the arrangement of the excerpts is designed to prove a Marxist-Leninist-Maoist point. The data, however, can be used independently of the framework in which they are here placed; and the sources of all the excerpts are clearly noted, thus allowing this volume (and the others in the series) to be used as an index to the large corpus of sources from which it is compiled. The excerpts are arranged first chronologically, and then within large periods by subject. Among the subjects covered are: the Ch'ing land system, concentration of land ownership, rural usury, the Taiping land and tax systems, expansion of the cultivated area, exploitative tenancy relations, the land tax, salt tax, likin, corvée, the enlargement of the domestic and foreign market for agricultural products, the decline of rural handicrafts, the commercialization of agriculture, the state of agricultural technology, natural disasters, Ch'ing government policy toward agriculture, peasant discontent and uprisings. There is no introductory essay such as is found in the comparable volumes on modern industry (4.4.1-2). (2,600)

4.2.4 Chang Yu-i 章有义, comp., Chung-kuo chin-tai nung-yeh shih tzu-liao, ti-erh chi, 1912-1927 中國近代农业史資料第二輯 (Source materials on China's modern agricultural history, second collection, 1912-1927; Peking: San-lien shu-tien, 1957), 16 + 746 pp.

A continuation of the previous item for the years of the Warlord Period and the 1927 Nationalist Revolution, with the excerpts drawn largely from periodicals, agricultural reports, and specialized studies. A bibliography of sources is included (pp. 735-746). (2,600)

4.2.5 Chang Yu-i 章有义, comp., Chung-kuo chin-tai nung-yeh shih tzu-liao, ti-san chi, 1927-1937 中国近代农业史資料第三輯 (Source materials on China's modern agricultural history, third collection, 1927-1937; Peking: San-lien shu-tien, 1957), 16 + 1077 pp.

A continuation of the two previous items for the Kuomintang period to 1937, designed to demonstrate the decay of China's rural economy as a result of "feudal compradore" and "imperialist" exploitation. The agricultural reform movements of, e.g., Liang Shu-ming and the CCP-led peasant movement also receive some attention. These very important excerpts have been drawn from several hundred agricultural reports, periodicals, and specialized studies which are listed on pages 1066-77. (2,600)

4.3 THE "CAPITALIST BURGEONS" CONTROVERSY

Between 1955 and 1957 one of the most controversial topics in Chinese Communist historiography was the problem of the origins of capitalism in China. If it were asserted, as for example in the quotation by the historian Li Shu cited in sec. 2.4 above, that China's "feudal" economy was developing and changing, however slowly, then that economy should naturally have developed toward capitalism--if one follows the Marxist scheme of things. In fact, Mao himself supplied the basic text on this subject which aroused considerable exegetical writing between 1955 and 1957: "As China's feudal society developed its commodity economy and so carried within itself the embryo of capitalism, China would of herself have developed slowly into a capitalist society even if there had been no influence of foreign imperialism" (Selected Works [English ed., London, 1954], III, 77). Although the exegetes agreed that China's feudalism was developing toward capitalism, they differed quite sharply as to the date which they assigned to the origins of capitalist burgeons (the word used is *meng-ya* 萌芽, "sprouts," "shoots," "burgeons"), and also in the degree of development which they postulated. For a time, during 1956 and 1957, it appeared that the view represented by the historian Shang Yüeh had won the day. In brief, Shang and his supporters argued that the late Ming and Ch'ing economy was already proto-capitalist. The central arch of this contention was the assertion of the widespread existence of factory handicrafts (*kung-ch'ang shou-kung-yeh* 工廠手工業), which are presumed to have fulfilled the Marxian criteria for capitalist production. Their development, so the schema goes, was preceded by a proliferation of internal and external trade. These market forces acted to bring about an increasing differentiation of handicraft, traditionally a peasant ancillary occupation, from agriculture, as well as an unprecedented concentration of landholding, which forced many peasants into newly-growing towns where they found employment in factory handicraft. The new "bourgeoisie" of the late Ming (whose ideological leaders, it is explained, were such men as Ku Yen-wu, Huang Tsung-hsi, and Wang Fu-chih) would eventually have seized political power in combination with their peasant allies (this is the interpretation given to Li Tzu-ch'eng's rebellion which overthrew the Ming dynasty) and then proceeded to prepare the rest of the prerequisites for the development of industrial capitalism, just as their English and French counterparts are alleged to have done. But the Manchu invasion and devastation of the land in the first instance, and the imperialist aggression and exploitation which followed in the nineteenth century prevented this happy fruition.

Like the debates on the nature of Chinese society of the 1920's and 1930's, in which most of the theoretical arguments now advanced were already stated, the present concern to establish that China's premodern economy was in fact evolving in accordance with the Marxist normative stages of societal development can best be understood as an effort to claim an identical pedigree with the West, parallel and not derivative, and of equal hoariness. To assert that Chinese society was not fundamentally different from that of Europe and to ascribe the humiliations endured in the past century to the conspiracy of the Manchu Dynasty and its "compradore-feudal" successors with the imperialist powers, is doubtless a means by which the positive value of China's history may be maintained.

It may well be that we have until now underestimated the degree of commercialization of the premodern Chinese economy. But the step from the posited existence of extensive commerce and advanced forms of organization in handicraft manufacture to the assertion that the Chinese economy was developing toward an "industrial revolution" is an act of faith, rather than a "scientific" historical conclusion. Even more significant than the external criticism that we might apply is the fact that more recently the weight of authoritative opinion in Communist China seems to have swung away from Shang Yüeh.

Assuming that somewhere in the writings of Mao (or Marx, Engels, Lenin, and Stalin) there is the authoritative word, which of the incomplete and frequently offhand directives provided by the "classics of Marxism" shall be followed? In this instance, the remark by Mao just cited, beginning "As China's feudal society developed its commodity economy...," follows immediately after a passage which reads, "Chinese feudal society lasted for about three thousand years. It was not until the middle of the nineteenth century that great internal changes took place in China as a result of the penetration of foreign capitalism." Shang Yüeh's critics explicitly indicted him for contradicting this last passage. Behind their charge was the fear that too great an emphasis on internal proto-capitalist developments prior to the full impact of Western imperialism in the nineteenth century might divert attention from the villain's role assigned to foreign capitalism in transforming China into a semi-colonial, semi-feudal status. This clearly would not fit in with the need, at this stage of the Chinese revolution, to project the largest share of the blame for a century and more of humiliation and weakness onto the "imperialist aggressors." Thus Shang was accused of misinterpreting the nature of the Opium War, of failing to see that it marked the beginning of "the struggle between the Chinese people's anti-imperialist, anti-feudal line which sought to transform China into an independent and prosperous nation, and the imperialist feudal line which sought to transform China into a colony." This was because of his false attribution of a connection between the leadership of the "people's struggles" in 1839-42 and the "bourgeois-urban movement" of the late Ming and early Ch'ing, with the result that "the nature of this [the alleged popular resistance to the British during the Opium War] Chinese national anti-aggressive struggle is changed into 'bourgeois' anti-feudalism." Moreover, playing up the degree of China's economic development along the Marxist normative road to capitalism may raise doubts about the historical necessity of the revolution led by the Chinese Communist Party. "If three hundred years ago capitalism already held such a secure position," stated the critics of Shang Yüeh, "then the anti-feudal land reform led by the Communist Party could not have occurred....And how could there have been any necessity for the proletariat to seize the leadership of the democratic movement?" (See Li-shih yen-chiu [Historical studies], pp. 1-16 [Jan. 1958]; pp. 1-11 [Mar. 1959]).

In sum, then, the mainland historians have not been able to reconcile the need to value their own past--China's history--by demonstrating that it was following a path of development parallel to, but independent of, Western Europe, with the equally pressing need to show that it was the conscienceless aggression of the Western powers which undermined traditional Chinese society and directly caused the international humiliations and domestic crises of the century that began with the Opium War. Item 4.3.1 reprints a cross-section of the huge outpouring of articles devoted to the problem of the incipiency of capitalism in China. The next two items set forth the position taken by Professor Shang Yüeh. Items 4.3.6 and 4.3.7 demonstrate how "old-style" history and historians have also become involved in the controversy. The last two items deal with Ming overseas trade and its contribution to the development of incipient capitalism.

4.3.1 Chinese History Seminar, Chinese People's University, ed., Chung-kuo tzu-pen chu-i meng-ya wen-t'i t'ao-lun chi 中国資本主义萌芽問題討論集 (Collected papers on the problem of the incipiency of capitalism in China; Peking: San-lien shu-tien, 1957), 2 vols., 8 + 1102 pp.

Collected papers on the problem of incipient capitalism in China, consisting of reprints of thirty-three items from journals, newspapers, and pamphlets originally published between 1955 and 1957 and including ten from the leading historical publication, Li-shih yen-chiu. These articles seek to document the existence of a "capitalist" embryo within the womb of China's "feudal" society, to determine the level of its maturation before it was aborted, and to explain the slowness of its growth and the causes of its eventual demise. While a number of the contributors would go back to the late T'ang or the Sung, the majority hold that by the sixteenth century, the Wan-li reign of the Ming dynasty, capitalist meng-ya were flourishing in many handicraft industries and to some degree in agriculture as well. An impressive amount of documentation is offered in support of this thesis, but the argumentation never gets beyond filling in the proper Marxian boxes. The failure of these Ming beginnings to blossom into full-blown capitalism is laid principally to the harshness of the conquest by the Manchus, who entered China originally at the behest of the desperate Ming feudal remnants after the "bourgeois" urban movement and the "peasant" rebellion of Li Tzu-ch'eng 李自成 had coalesced to overthrow the Ming dynasty. The new flourishing of incipient capitalism and peasant unrest in the eighteenth and early nineteenth centuries might have broken the fetters of the "feudal" rule of the Ch'ing, if the traitorous dynasty had not joined with the foreign imperialists to defeat the Taipings and so prolong the life of the feudal regime. (12,000)

4.3.2 Shang Yüeh 尚鉞, Chung-kuo tzu-pen-chu-i kuan-hsi fa-sheng chi yen-pien ti ch'u-pu yen-chiu 中国資本主义关系發生及演变的初步研究 (Preliminary investigations of the origin and development of capitalist relations in China; Peking: San-lien shu-tien, 1956), 11 + 280 pp.

Three long articles by Professor Shang Yüeh of the Chinese People's University in Peking in which he sets forth in detail his position in the "capitalist burgeons" controversy: "The sprouting and growth of capitalist factors of production in China" (pp. 1-72); "Stagnation, change, and development of Chinese society in the early Ch'ing period" (pp. 73-144--this paper also is printed in Chung-kuo tzu-pen chu-i meng-ya wen-t'i t'ao-lun chi, item 4.3.1); and "The development and change of scholarly thought in the late Ming and early Ch'ing" (pp. 145-271). The articles are heavily bolstered by quotations from Chinese sources and the Marxist classics. Item 4.3.3 is a companion volume of essays by Shang Yüeh's students. (10,000)

4.3.3 Chinese History Seminar, Chinese People's University, comp., Ming Ch'ing she-hui ching-chi hsing-t'ai ti yen-chiu 明清社会經済形態的研究 (Studies on the social and economic structure of the Ming and Ch'ing; Shanghai: Shang-hai jen-min ch'u-pan-she, 1957), 7 + 357 pp.

This volume consists of four student dissertations written at the People's University in 1955, entitled: (1) "The development of a commodity economy and the beginnings of capitalism in the Ming period"; (2) "The origin and function of the 'Single Whip' system in the Ming period"; (3) "An investigation into the question of the Ming domestic market"; (4) "On the development of commercial agriculture in the Ch'ing before the Opium War." These studies were apparently produced under the direction of Professor Shang Yüeh, who contributes a preface. They all support his general thesis (see 4.3.2) concerning the relatively high stage of development of capitalist "burgeons" in late

Ming and early Ch'ing China and the consequent progressive decay of the feudal economy even prior to the impact of Western capitalism. (18,000)

4.3.4 Ch'ien Hung 錢宏, Ya-p'ien chan-cheng i-ch'ien Chung-kuo jo-kan shou-kung-yen pu-men chung ti tzu-pen-chu-i meng-ya 鴉片戰爭以前中国若干手工部門中的資本主义萌芽 (Incipient capitalism in several sectors of Chinese handicraft industry prior to the Opium War; Shanghai: Shang-hai jen-min ch'u-pan-she, 1956), 42 pp.

Silk and cotton manufacture in southeast China, pottery at Ching-te-chen 景德鎮, iron implements at Fo-shan 佛山 in Kwangtung, and metallurgy, lumbering and paper manufacture in Shensi--i.e., the usual examples cited in most works on this subject--are briefly described in such a way as to support the author's contention that incipient capitalist elements were present in these handicraft industries from the Ming dynasty on, and that the feudal property system prevented the further development of capitalism. This paper is also printed in 4.3.1, pp. 238-271. (13,000)

4.3.5 Fu Chu-fu 傅筑夫 and Li Ching-neng 李競能, Chung-kuo feng-chien she-hui nei tzu-pen chu-i yin-su ti meng-ya 中国封建社会内資本主义因素的萌芽 (Incipient capitalistic elements in Chinese feudal society; Shanghai: Shang-hai jen-min ch'u-pan-she, 1956), 53 pp. (15,000)

4.3.6 Fu I-ling 傅衣凌, Ming Ch'ing shih-tai shang-jen chi shang-yeh tzu-pen 明清时代商人及商業資本 (Merchants and commercial capital in the Ming and Ch'ing periods; Peking: Jen-min ch'u-pan-she, 1956), 216 pp.

Since this book deals with Ming and early Ch'ing, its title is not very precise, as the author (a professor at Amoy University) states in his postface. The seven chapters (one a general account of the subject, four on the Ming and two on the Ch'ing) are in fact a group of essays which consider the rôle of merchants and merchant capital in several key areas (Hui-chou , Anhwei; Tung-t'ing , Kiangsu; Fukien; Shensi; the maritime copper trade; foreign trade at Amoy) and their effect on Chinese "feudal" society. In style, the essays consist in large part of extensive quotations strung together by the author's commentary, reflecting the fact that, except for the first, they were originally written in the "old style" prior to 1949. The footnotes to each chapter indicate the use of extensive contemporary sources, but there is no bibliography. (6,000)

4.3.7 Fu I-ling 傅衣凌, Ming-tai Chiang-nan shih-min ching-chi shih-t'an 明代江南市民經济試探 (An exploration into the urban economy in the Kiangnan area during the Ming dynasty; Shanghai: Shang-hai jen-min ch'u-pan-she, 1957), 131 pp.

The author believes that the years 1522-66 marked the beginning of "capitalist burgeons" in China, and that the region south of the Yangtze and along the coast was the first to manifest them. This is the thesis that ties together his five heavily-documented chapters on: (1) the use of silver and increased circulation of commodities in this period; (2) the economy of the "wealthy households" (landlords, merchants, and handicraft producers); (3) new developments in the landlord economy; (4) the textile industry and the textile workers' uprisings; and (5) the "anti-feudal movements" of the lower classes in Kiangnan in the late Ming. An appendix discusses cotton-cloth firms in the Ming and Ch'ing periods. (4,000)

4.3.8 Chang Wei-hua 張維華, Ming-tai hai-wai mao-i chien lun 明代海外貿易簡論 (A brief discourse on overseas trade in the Ming dynasty; Shanghai: Shang-hai jen-min ch'u-pan-she, 1956), 2 + 115 pp.

This study deals with: (1) the Ming government's restrictive policies toward private overseas trading in the interest of establishing or maintaining tribute-type relations; (2) the ways in which private overseas trading developed, in spite of the Ming policies and notwithstanding the disruptive effect of Japanese and other piracy (includes a table [pp. 58-72] listing tribute goods and major items of commerce of several countries with which the Ming empire traded, based principally on Ming hui-tien 明會典); (3) relations between China and the Nanyang countries, and the impact of Chinese trade upon the latter. The author sees the development of the Chinese economy in this period in terms of "incipient capitalism," while he praises the resistance of the overseas Chinese and local Nanyang peoples to European colonialism. (2,000)

4.3.9 T'ien Ju-k'ang 田汝康, 17-19 shih-chi chung-yeh Chung-kuo fan-ch'uan tsai Tung-nan ya-chou 17-19世紀中業中国帆船在東南亞洲 (Chinese sailing ships in Southeast Asia from the seventeenth to the mid-nineteenth century; Shanghai: Shang-hai jen-min ch'u-pan-she, 1957), 45 pp. (5,000)

4.4 INDUSTRY AND HANDICRAFT

Chinese Communist historiography distinguishes three types of modern industry in late nineteenth and twentieth century China: foreign-owned (or "imperialist") enterprises, government-sponsored and semi-official undertakings such as the kuan-tu shang-pan 官督商辦 ("official supervision and merchant management") firms organized by Sheng Hsuan-huai, and third, Chinese-owned private capitalist enterprises. It is argued that foreign investment in manufacturing in the treaty ports after 1895, along with the vast flood of foreign manufactured imports, acted to undermine the market for traditional Chinese handicraft production, especially for the products of the ancillary spinning and weaving activities carried on by the peasantry in off seasons. Imported goods and foreign manufactures in the treaty port enclaves created a demand in China for new products, and simultaneously, because of their higher quality and lower price, they drove competing Chinese handicraft manufactures out of the market. But because of the inability of foreign manufactured goods immediately to supply the new demand, part of the market for modern machine-made commodities was left to be exploited by pioneer Chinese industrial firms. It was in this small niche created by imperialism that modern industry in China is considered to have had its beginnings.

The first modern Chinese industrial enterprises were government-sponsored arsenals, shipyards, and the like. These were followed by official or semi-official mining, metallurgical, and textile firms. This official industrialization effort is looked upon rather equivocally by the Chinese Communist historians, as we have seen in sec. 2.7 in connection with the late nineteenth century "Westernization movement." One of the chief purposes of the industrialization efforts by Li Hung-chang and others was to provide support to the faltering Manchu "feudal" regime. Their enterprises were the forerunners of the "compradore capitalism" of the Kuomintang period. In contrast, praise goes to the "private" capitalist firms which, struggling against the monopolistic position of the semi-official undertakings, yet managed to survive. Their promoters and managers are seen as the forerunners of the "national bourgeoisie" who, in the ideology of the People's Republic of China, were part of the coalition which overthrew the compradore-feudal Kuomintang regime. Because Marx had stated that the "really revolutionary way" in which industrial capitalism develops (as distinguished from merchant investment in manufacturing) is for a section of the craftsmen themselves to accumulate capital and become merchants and capitalists, evidence of this process has also been assiduously sought by the hardworking compilers of documentary materials.

The truth is, of course, that there was very little modern industry in China prior to 1949 in any of the three categories proposed by the Chinese Communist historians. In order, however, to have a proletariat--and a proletariat is a sine qua non before there can be a Communist Party--there must first be modern industry. This perhaps accounts for the tremendous energies expended in compiling 4.4.1 and 4.4.2. The introductory essays to these two collections of source materials together form a very useful brief history of modern industry in China. Items 4.4.3-6 deal with the history of several individual industries and include an excellent study by Yen Chung-p'ing, who is probably the most capable of the economic historians now writing in Communist China. The career of the chief engineer of the first railroad in China built without the aid of foreign funds is treated in 4.4.7 and 4.4.8. While "imperialist" investment and foreign firms in China are treated in most of the preceding works as well, they are the special subject of 4.4.9-12. Item 4.4.13 is an immense collection of source materials on handicraft industry in modern China; and the last three items, including Yen Chung-p'ing's study of copper mining in Yunnan, are also concerned with premodern industrial undertakings. See also 4.1.1.

4.4.1 Sun Yü-t'ang 孫毓棠, comp., Chung-kuo chin-tai kung-yeh shih tzu-liao, ti-i chi, 1840-1895 nien 中国近代工业史資料第一輯 1840-1895 年 (Source materials on the history of modern industry in China, first collection, 1840-1895; Peking: K'o-hsueh ch'u-pan-she, 1957), 2 vols., 77 + 1271 pp., plates.

An enormous compilation of excerpts from Western books and periodicals, Ch'ing documentary collections, and (to a lesser extent) archival materials concerning China's nineteenth-century "industrialization." Part I (pp. 3-248) draws largely on the North China Herald to provide data on about 103 foreign-owned industrial enterprises; Part II (pp. 249-566) treats government-sponsored arsenals, shipyards and the like; Part III (pp. 567-956), official mining, metallurgical, and textile enterprises; Part IV (pp. 957-1173), enterprises undertaken by private capital--more than 100; and Part V (pp. 1194-1254) includes data on the labor force, working conditions, and wages. The sources of all excerpts are carefully indicated. The whole is preceded by a sixty-six page introductory essay which is, in effect, a brief history of nineteenth-century Chinese industry. There are, in addition, a bibliography of works from which the materials were taken and a table of names of foreign firms and persons with their Chinese equivalents. The stupendous labor involved in producing such a collection could not easily be duplicated outside of China. This (and the other collections of this type: see 4.2.3-5, 4.4.2, and 4.4.13) will necessarily be a starting point for all researchers on China's modern economic history, and an index to the vast body of sources that must be tapped. (6,975)

4.4.2 Wang Ching-yü 汪敬虞, comp., Chung-kuo chin-tai kung-yeh shih tzu-liao, ti-erh chi, 1895-1914 nien 中国近代工业資料史第二輯 1895-1914年 (Source materials on the history of modern industry in China, second collection, 1895-1914; Peking: K'o-hsueh ch'u-pan-she, 1957), 2 vols., 67 + 1329 pp., plates.

A continuation of the previous item, also consisting of an impressive mass of excerpts from Chinese, Japanese, and Western sources, illustrating the history of modern industry in China from the Sino-Japanese War to the eve of World War I. The compiler's fifty-page introduction, which strings the quoted materials together on a Marxist-Leninist thread, provides a compact industrial history of this period. Part I (pp. 1-411) treats foreign investment and foreign firms in China, 1895-1913, and, like the succeeding parts, is organized topically, industry by industry; Part II (pp. 412-647), government and semi-official industrial enterprises; Part III (pp. 649-920), the beginning of private industry; Part IV (pp. 921-1171), the financing of private enterprise and the market for its products; anc Part V (pp. 1172-1301), the labor force, wages and working conditions, and strikes. A bibliography of works excerpted, and tables of Chinese and Western equivalents for the names of persons and firms are appended. These materials are invaluable, have often been taken from very rare sources, and seem to be quoted without alteration of the original texts. However, because they are often only brief excerpts, the editor has had little trouble in arranging them under tendentious and often misleading headings. (6,975)

4.4.3 Yen Chung-p'ing 嚴中平, Chung-kuo mien fang-chih shih kao 中国棉紡織史稿 (A draft history of Chinese cotton-spinning and waving; Peking: K'o-hsueh ch'u-pan-she, 1955), 5 + 384 pp., many tables.

This is a revised and enlarged version of Yen's Chung-kuo mien-yeh chih fa-chan 中国棉業之發展 (The development of China's cotton industry) which appeared on very poor paper in Chungking in 1942. The period covered is 1289-1937, but the centuries prior to the Opium War are disposed of in the first fifty-two pages. The succeeding chapters represent a detailed, scholarly, and readable history of the textile industry in China, treating in turn the impact of foreign cotton on Chinese handicraft production, the beginnings of modern textile manufacture by Chinese and foreign firms, its expansion during and after World War I, the effects of the depression, and the transformation of the handicraft textile industry. Two appendices give short sketches of the histories of the principal Chinese and foreign firms, 1890-1937 (pp. 341-367), and twenty-one statistical tables (in addition to numerous other tables in the text) on many aspects of the textile industry (pp. 368-384). The theoretical framework is, of course, Marxist, and much attention, for example, is given throughout to the problem of "national capitalism" in the face of foreign competition and imperialist aggression. But, like Yen's study of the Yunnan copper industry (4.4.16), the scholarly character and comprehensiveness of the work make it the best book that has yet appeared on its subject. (2,457)

4.4.4 Shanghai Economic Research Institute, Chinese Academy of Sciences, and the Economic Research Institute, Shanghai Academy of

Social Sciences, comps., Nan-yang hsiung-ti yen-ts'ao kung-ssu shih-liao 南洋兄弟烟草公司史料 (Historical materials concerning the Nanyang Brothers Tobacco Company; Shanghai: Shang-hai jen-min ch'u-pan-she, 1958), 31 + 756 pp., plates.

Documents from the archives and accounts of the Nanyang Brothers Tobacco Co., Ltd., and supplementary materials from private letters, personal interviews, memoirs, newspapers and the like, illustrative of the history of this Shanghai firm from its foundation by overseas Chinese entrepreneurs in Hong Kong in 1905 down to 1957. Very little detailed documentation on the internal history of industrial enterprises in China has hitherto been available. These materials are therefore of major value even though they represent only a part of the company's archives and the selection and arrangement have obviously been made to prove a point, i.e., that the development of "national capitalist industry" in a "semi-feudal, semi-colonial country" is subject to political and economic obstacles and that, in contrast, all is well after "liberation" and transformation into a "joint public-private" enterprise. The documents are divided into four periods (1905-19, 1919-36, 1937-49, 1949-57); within each period the treatment is topical but in a rough chronological order. In general the data on the early history of the company, prior to 1919, are scanty; and there are gaps for the Shanghai operations before 1937 due to the burning of the Shanghai plant and archives during the Japanese attack of that year. The source of each excerpt is carefully indicated, and the editors have added headings, punctuation, and brief notes where necessary. (1,000)

4.4.5 Shanghai Economic Research Institute, Chinese Academy of Sciences, and the Economic Research Institute, Shanghai Academy of Social Sciences, comps., Ta-lung chi-ch'i ch'ang ti fa-sheng fa-chan yü kai-tsao 大隆机器厂的发生发展与改造 (The origin, development, and transformation of the Ta-lung Engineering Company ["Oriental Engineering Works, Ltd."]; Shanghai: Shang-hai jen-min ch'u-pan-she, 1959), 6 + 124 pp.

A brief history of one of the earliest private, Chinese-owned engineering enterprises in Shanghai. The "Oriental Engineering Works, Ltd." was founded in 1902 by one Yen Yü-t'ang 严裕棠, and at first was principally engaged in the repair of steamer and textile machinery. The firm gradually began to manufacture machinery and parts itself, and just before the Sino-Japanese War of 1937 it put on the market complete cotton spinning equipment. After 1949, the company became a joint public-private enterprise. The argument of the present volume centers on contrasting the obstacles faced by the firm and the hardships endured by its employees before "liberation" with its considerable expansion and flourishing under the new regime. Considerable reliance has been placed on interviews with veteran officers and employees. (2,900)

4.4.6 Shanghai Economic Research Institute, Chinese Academy of Sciences, and the Economic Research Institute, Shanghai Academy of Social Sciences, comps., Heng-feng sha-ch'ang ti fa-sheng fa-chan yü kai-tsao 恒丰纱厂的发生发展与改造 (The origin, development, and transformation of the Heng-Feng Spinning Mill; Shanghai: Shang-hai jen-min ch'u-pan-she, 1959), 7 + 152 pp.

The Heng-feng Mill, dating from 1891 and therefore one of the earliest in China, began as a kuan-tu shang-pan enterprise under the aegis of Li Hung-chang and Sheng Hsuan-huai 盛宣懷. It soon passed into the hands of Nieh Chi-kuei 聶緝槼, a son-in-law of Tseng Kuo-fan, and remained under the control of the Nieh family until it became a joint public-private enterprise after 1949. This is a short but quite detailed history of the company, its management, and its operations, based on the firm's records (which, however, are very fragmentary), and on interviews with management and workers. Pages 121-152 deal with the period 1949-57, and are offered in contrast to the company's pre-1949 vicissitudes. (3,200)

4.4.7 Hsü Ying 徐盈, Li Hsi-mi 李希泌, and Hsü Ch'i-heng 徐啟恒, Chan T'ien-yu 詹天佑 (A biography of Chan T'ien-yu; Peking: Chung-kuo ch'ing-nien ch'u-pan-she, 1956), 54 pp., 6 plates. (20,000)

4.4.8 Hsü Ch'i-heng 徐啟恒 and Li Hsi-mi 李希泌, Chan T'ien-yu ho Chung-kuo t'ieh-lu 詹天佑和中国铁路 (Chan T'ien-yu and Chinese railroads; Shanghai: Shang-hai jen-min ch'u-pan-she, 1957), 2 + 92 pp., 4 plates, 2 maps.

Chan (1861-1919) was a graduate of Yale, having come to the United States as part of the Chinese Educational Mission; he became the chief engineer of the first Chinese railroad (Peking-Kalgan) built without foreign financial help. (7,500)

4.4.9 Wei Tzu-ch'u 魏子初, ed., Ti-kuo chu-i yü K'ai-lan mei-k'uang 帝国主义与開灤煤礦 (Imperialism and the Kailan mines; Shanghai: Shen-chou kuo-kuang she, 1954), 23 + 230 pp.

Aside from a fourteen-page introduction which describes the development of the Kailan mines in terms of British imperialist treachery and collusion with Chinese bureaucratic capitalists, this volume consists of documentary materials with notes by the editor. Arranged under eleven headings, the documents include: (1) memorials by Li Hung-chang, Yuan Shih-k'ai, and lesser Chinese officials concerned with the Kaiping and Lanchou mines between 1881 and 1911; (2) agreements for the transfer of the Kaiping Mining Company to the Oriental Syndicate (1910); (3) records of the lawsuit (1905) of Chang I (also known as Chang Yen-mao , Chinese director of the company) against C. Algernon Moreing, British engineer and promoter who had been represented by Herbert Hoover in the purchase of the mines; (4) proclamations, regulations, and the like, relating to the establishment of the Lanchou mines (1907); (5) materials related to the formation of the Kailan Mining Administration (1912); (6) various papers related to the area in Chinwangtao sold to the Oriental Syndicate along with the Kaiping mines. Preceding the text is a brief discussion of source materials on the Kailan mines. Discrepancies between the English and Chinese versions of the various documents are pointed out by the editor in footnotes. These materials should be of use in connection with Ellsworth Carlson's study, <u>The Kaiping Mines (1877-1912)</u> (Harvard University Press, 1957). (5,000)

4.4.10 Ch'en Chen 陳真 et al., comps., <u>Chung-kuo chin-tai kung-yeh shih tzu-liao, Ti-erh chi, Ti-kuo chu-i tui Chung-kuo kung k'uang shih-yeh ti ch'in-lueh ho lung-tuan</u> 中国近代工业史資料, 第二輯, 帝国主义对中国工矿事业的侵略和壟斷 (Source materials on China's modern industrial history, second collection: Imperialist aggression against and monopolization of China's industries and mines; Peking: San-lien shu-tien, 1958), 7 + 975 pp.

Selections from newspaper and periodical articles, books, company reports, Japanese wartime and Chinese post-1949 field investigations, and archival materials illustrating the history of foreign industrial and mining firms in China. This vast storehouse of material, tendentiously chosen and arranged, consists first of seven chapters dealing respectively with England (pp. 1-266), the United States (pp. 267-386), Japan (pp. 387-715), France (pp. 717-754), Germany (pp. 755-788), Tsarist Russia (pp. 789-807), and other countries (pp. 809-835). For each country the presentation of general data on "imperialist" investment is followed by a firm-by-firm round-up material. The two final chapters contain statistical materials on profits and "surplus value" (pp. 837-949), and tables showing the relative positions of "Chinese industry" and "imperialist industry in China" (pp. 951-975). In general, the data are most extensive for the 1920's and 1930's; many statistical tables and charts are included; and the sources drawn upon are in Chinese, English, and Japanese, and are cited at the end of each section. The editors include frequent explanatory notes in smaller type, but there is no introduction to the volume as a whole. (3,150)

4.4.11 Wei Tzu-ch'u 魏子初, Ti-kuo chu-i tsai Hua t'ou-tzu 帝国主义在華投資 (Imperialist investments in China; Peking: Jen-min ch'u-pan-she, 1951), 25 pp., 5 tables. (8,000)

4.4.12 Wu Ch'eng-ming 吴承明, Ti-kuo chu-i tsai chiu Chung-kuo ti t'ou-tzu 帝国主义在旧中国的投資 (Imperialist investments in the old China; Peking: Jen-min ch'u-pan-she, 1955), 2 + 188 pp. (3,500)

4.4.13 P'eng Tse-i 彭澤益, comp., Chung-kuo chin-tai shou-kung-yeh shih tzu-liao (1840-1949) 中国近代手工业史資料 (Source materials on the history of the handicraft industry in modern China [1840-1949]; Peking: San-lien shu-tien, 1957), 4 vols., ca. 2800 pp., plates and tables.

The materials in this huge collection of excerpts are drawn from nearly 1,000 Chinese, Japanese, and Western-language sources (including 256 local gazetteers [ti-fang-chih], numerous Ch'ing official and private collections, monographs, reports of field investigations, newspapers and periodicals, Imperial Maritime Customs publications, and travelers' accounts) which are listed on pp. 561-598 of Vol. 4. The excerpts are first grouped chronologically into six periods, and within each period are arranged by topics: (1) The mid-seventeenth century to 1840: handicraft technology, state handicraft industry, private handicraft industry and the gild system, the beginning of capitalist handicraft production. (2) 1840-70: the influence of the treaty ports, Taiping state handicraft industry, Ch'ing state handicraft production of arms and copper cash. (3) 1870-1900: handicraft vs. modern-type industry, stimulation of handicraft output resulting from an enlarged market, decline and stagnation. (4) 1900-20: development of handicrafts near the treaty ports, competition from foreign goods, Ch'ing efforts to encourage technology, labor unrest among handicraft workers, the effects of World War I. (5) 1920-30: development of domestic capitalism, workers' conditions, effects of the depression, the structure of handicraft production. (6) 1930-49: Japanese oppression of handicraft in the occupied areas, the crisis of handicraft production in the KMT areas, U.S. imperialist and Chinese bureaucratic joint damage to handicrafts. From the above, it is obvious that the topical arrangement is "loaded." But the materials remain valuable in themselves and, as in the other volumes in this series, the selections here printed, with sources fully cited, may be used as an index to the huge literature from which they are drawn. There is no introductory essay such as is found in the volumes on modern industry. (2,500)

4.4.14 Ch'en Shih-ch'i 陳詩啟, Ming-tai kuan shou-kung yeh ti yen-chiu 明代官手工業的研究 (A study of the official handicraft industry in the Ming dynasty; Wuhan: Hu-pei jen-min ch'u-pan-she, 1958), 2 + 183 pp. (2,600)

4.4.15 Pottery and Porcelain Research Institute, Bureau of Light Industry, Kiangsi Province, comp., Ching-te-chen t'ao tz'u shih kao 景德鎮陶瓷史稿 (Draft history of the pottery and porcelain of Ching-te-chen; Peking: San-lien shu-tien, 1959), 5 + 434 pp., 130 illus.

This is a comprehensive, well-documented, and amply-illustrated history of Chinese pottery and porcelain as produced at the famous Ching-te-chen kilns from roughly the first century A.D. to the present. The historical stages of the ceramics industry are arranged according to the generally accepted Communist periodization scheme, i.e., the feudal period (up to the Ming, pp. 43-94), the period of incipient capitalism (Ming and Ch'ing, pp. 95-255), the semi-colonial, semi-feudal society (from the Opium War to 1949, pp. 259-346), and New China (pp. 349-434). Within this general outline, there is considerable substantive material regarding the sites, techniques of pottery manufacture, the specific types of products (especially those of the Ming and Ch'ing), the porcelain market, and the employment of workers. Obviously anxious to point up improvements under the Communist regime, the book nonetheless ends on a cautious note about the problems faced by the industry, e.g., due to competition from Japan. (2,000)

4.4.16 Yen Chung-p'ing 嚴中平, Ch'ing-tai Yun-nan t'ung cheng k'ao 清代雲南銅政考 (A study of the Yunnan copper administration in the Ch'ing period; Peking: Chung-hua shu-chü, 1957), 4 + 100 pp.

This volume is a clear and objective monograph on the rise and fall of the Yunnan copper industry in the Ch'ing period. It examines the problems of the production and transport of copper and the minting of copper cash, and it includes an account of the kuan-tu shang-pan 官督商辦 ("official supervision and merchant management") mining company which was formed in 1884 in an effort to revive the industry after the mid-nineteenth-century rebellions. In addition to surveying the history of this enterprise, the author describes the forms of organization and the production methods used by the Ch'ing government and the company and their various inadequacies. An appendix gives statistical tables (pp. 79-100) on copper production, sales, minting of cash, and the like. Sources are given in 119 reference notes. (1,900)

4.5 MONEY, BANKING AND PRICES

It should be evident to the reader by now that comprehensive and reliable statistics on any aspect of China's economy prior to 1949 are scarce indeed. In the important area of prices, 4.5.5-7 are among the most valuable published materials available. A comprehensive history of modern banking in China will have to be based on the archives of both the foreign and Chinese establishments which grew up in the treaty ports in the last part of the nineteenth century and the twentieth century. Unfortunately, neither of these sources is yet available to the scholar. Items 4.5.3 and 4.5.4 are brief reconnaissances that are far from filling the need in this area. Item 4.5.1 will surely be a standard survey of the history of Chinese currency for some time to come.

4.5.1 P'eng Hsin-wei 彭信威, Chung-kuo huo-pi shih 中國貨幣史 (A history of Chinese currency), rev. ed. (Shanghai: Shang-hai jen min ch'u-pan-she, 1958), 49 + 675 pp., 91 plates.

A comprehensive and systematic treatment of Chinese currency from the earliest stage down to the twentieth century (end of the Ch'ing period). For each dynasty or period, there are discussions of the several kinds of currency, their purchasing power, various theories concerning the currency, and credit and credit organizations. Extensive footnotes indicate wide-ranging use of sources in English, Japanese, and French, as well as in Chinese. Pages 429-520 concern the Ming period, pp. 521-675, the Ch'ing. This is a standard survey of the subject, and it is crammed with innumerable tables and data on all manner of financial topics, not merely currency. (1,200)

4.5.2 Cheng Chia-hsiang 鄭家相, Chung-kuo ku-tai huo-pi fa-chan shih 中国古代貨幣發展史 (A history of the development of currency in ancient China; Peking: San-lien shu-tien, 1958), 6 + 202 pp.

The fourteen chapters of this book deal with the evolution of currency fron pre-historic times through the Warring States period. The text is liberally supplied with illustrations of cowrie shells and ancient metal currency. Despite the author's claim to combining the roles of collector and student of history, the result is almost entirely descriptive. There are no footnotes or bibliography. (3,000)

4.5.3 Chang Yü-lan 張郁蘭, Chung-kuo yin-hang-yeh fa-chan shih 中国銀行業發展史 (A history of the development of banking in China; Shanghai: Shang-hai jen-min ch'u-pan-she, 1957), 144 pp.

This is a brief survey of modern banking in China from 1896 to 1937, based in large part on banking periodicals and, for the pre-Republican period, on the published papers of Sheng Hsüan-huai 盛宣怀 and others. It contains little new data and the "analysis" is forced. The author either has not had access to or has not attempted to use the voluminous bank archives that must have become available after 1949 and from which alone a full account of Chinese modern banking can be written. (2,000)

4.5.4 Hsien K'o 献可, Chin pai nien lai ti-kuo chu-i tsai Hua yin-hang fa-hsing chih-pi kai-k'uang 近百年来帝国主义在华銀行发行紙币概况 (A general account of the issuance of paper currency in China by imperialist banks in the past hundred years; Shanghai: Shang-hai jen-min ch'u-pan-she, 1958), 9 + 187 pp.

The purpose of this book, in the words of the author, is "to show something of the process of the colonization of the Chinese economy." To that end a vast amount of largely undigested statistical material (there are eighty-one tables) on what the author claims is a particularly critical aspect of the operation of the "imperialist banks" in China has been collected and arranged, with running commentary, into six chapters. These treat the introduction of foreign dollars into China; the banking activities in China during the past century of Britain, the United States, and Japan; paper currencies in the Japanese-occupied areas between 1937 and 1945; the issuance of paper currency by French, German, Dutch, and Belgian banks in China; and the paper currency activities of joint Sino-foreign banks. Much use is made of Japanese sources, but apparently none at all of company records and other unpublished materials that must surely have fallen into Communist hands during and after the civil war. (1,600)

4.5.5 Wu Kang 吳岡, comp., Chiu Chung-kuo t'ung-huo p'eng-chang shih liao 舊中國通貨膨脹史料 (Historical materials on the inflation in old China; Shanghai: Shang-hai jen-min ch'u-pan-she, 1958), 4 + 245 pp.

The "old China" in this book title starts with the 1935 "currency reform" by the Kuomintang government and ends with the Communist takeover of Shanghai on May 25, 1949. The currency system in this period, characterized by the compiler as "semi-feudal" and "semi-colonial," and its vicissitudes are described through selected excerpts from yearbooks, news reports, and journal articles. Some of the many statistical tables (which comprise most of the text from p. 140 on) are from unpublished sources including the files of the Central Bank of China. (4,000)

4.5.6 Institute of Economics, Nanking University, comp., 1913 nien - 1952 nien Nan-k'ai chih-shu tzu-liao hui-pien 1913-1952年南开指数資料汇編 (Nankai index numbers for 1913-1952; Peking: T'ung-chi ch'u-pan-she, 1958), 2 + 324 pp.

Some of the major statistical series of the Nankai Institute of Economics are here brought together and published in one volume: annual index of Tientsin wholesale prices, 1913-42; monthly index of Tientsin wholesale prices, 1928-52; annual and monthly index of the cost of living of Tientsin workers, 1926-42; monthly index of Tientsin workers' cost of living, 1946-49; weekly index of Tientsin workers' cost of living, 1950-52; Tientsin wholesale prices, 1913-42 (annual); Tientsin wholesale prices, 1928-52 (monthly); Tientsin retail prices, 1926-42 (annual); and Tientsin retail prices, 1926-52 (monthly). Because its statistical work was "bourgeois in method," the Institute's series were not continued after 1952. A brief introduction summarizes the work of the Nankai Institute and explains the methods used in compiling these statistics. (1,180)

4.5.7 Shanghai Economic Research Institute, Chinese Academy of Science, and the Economic Research Institute, Shanghai Academy of Social Sciences, eds., Shang-hai chieh-fang ch'ien-hou wu-chia tzu-liao hui-pien (1921-1957) 上海解放前后物价資料汇編 (A compendium of materials on commodity prices in Shanghai before and after "liberation"--1921-57; Shanghai: Shang-hai jen-min ch'u-pan-she, 1958), 12 + 600 pp.

Part I of this volume deals with Shanghai commodity prices in the thirty years before the Communist take-over, 1921-49. It is divided into four chapters (each containing numerous statistical tables and accompanying comment): (1) the overall situation of commodity prices, (2) wholesale commodity price indices, (3) wholesale prices of major commodities, and (4) workers' cost of living indices. Part II, dealing with the period June 1949 to 1957, contains three chapters and an appendix: (1) the overall situation of commodity prices,

(2) commodity price indices, and (3) major commodity prices. An appendix prints sixteen important government regulations and directives concerning commodity prices. Notwithstanding the political and ideological morals drawn by the editors in the foreword and the admittedly unsolved problems in this complex subject, this volume makes available a valuable body of apparently undoctored statistics, gathered from many different sources, which can provide the data for more refined analysis. (1,300)

4.5.8 Shou Chin-wen 寿進文, K'ang Jih chan-cheng shih-ch'i Kuo-min-tang t'ung-chih-ch'ü ti wu-chia wen-t'i 抗日战争时期国民党統治区的物价問題 (The question of commodity prices in the Kuomintang-controlled areas during the war against Japan; Shanghai: Shang-hai jen-min ch'u-pan-she, 1958), 4 + 78 pp. (3,500)

5. INTELLECTUAL AND CULTURAL HISTORY

This section is perhaps more uneven than others in this volume, consisting of two substantial sub-sections on intellectual history, followed by four others which are much smaller in bulk and in the case of the last two are merely suggestive of the considerably wider range of books available on their subjects. We continue to honor the practice of historians who begrudgingly add a chapter, or perhaps only a few footnotes, on intellectual and cultural matters to their large tomes of political, diplomatic, and economic history. In so doing we reflect not only our own shortcomings, but also those of the Chinese Communist historians. Much that one might expect to find included under the heading "intellectual and cultural history" is not being studied in the People's Republic of China. We have seen almost no works, for example, on the history of religion in modern China, on the history of education, on the evolution of the family, or on the history of the legal system. Even in the case of intellectual history proper, for that matter, the work of the past decade has tended to concentrate upon "progressive" thinkers, and with very few exceptions there has been little attention to what was after all the dominant stream of "conservative" or "reactionary" thought.

5.1 PRE-NINETEENTH-CENTURY THOUGHT

The intellectual history of any period, along with its art, literature, and legal system, are of course considered by Marxist writers to be part of the "superstructure" of the society in question, the form and content of which are directly dependent on the infrastructure of the "social relations of production." A feudal society produces feudal thought; a bourgeois society, bourgeois thought; and a socialist society, socialist thought. It is within this rigid framework that the Chinese Communist historians too interpret their intellectual history. Even from the Marxist point of view, however, it is possible for a given society, whether classified as feudal, bourgeois, or socialist, to contain niches which harbor antagonistic intellectual currents, due to the imperfect integration of any social system and the resultant omnipresent class struggle. Thus even in a slave or a feudal society, there may be proponents of "materialistic" thought who reflect the outlook of the oppressed classes (5.1.3). The aspirations of the oppressed may also find an outlet in the form of Utopian fantasies which represent a criticism of the existing class relations. But these aspirations, say the mainland historians, can never be realized (5.1.7-8), for only the "scientific" guidance of Marxism-Leninism can lead to a real utopia.

The current treatment of China's intellectual history consists, then, largely in connecting the writings of each thinker with his class background and with the contemporary configuration of class forces. Item 5.1.1 is the most comprehensive work of this kind that has appeared to date. Although the interpretation is mechanical, the extensive quotation of original sources makes Hou Wai-lu's history a useful reference tool. The mainland treatment of early modern intellectual history is closely tied in with the "capitalist burgeons" controversy already discussed in sec. 4.3. Here the effort is to show that the ideas of such men as Ku Yen-wu, Huang Tsung-hsi, and Wang Fu-chih reflect the appearance of the embryo of capitalism within the womb of Chinese feudal society. These men and their counterparts are now the objects of considerable study in mainland China (5.1.2, 5.1.9-14). Even when a bald assertion of the bourgeois nature of nonconformist thought is not made, special attention is paid to heretical figures of almost any variety as, for example, by republishing item 5.1.15.

5.1.1 Hou Wai-lu 侯外廬 et al., Chung-kuo ssu-hsiang t'ung-shih 中国思想通史 (A general history of Chinese thought), 4 vols.

(Peking: Jen-min ch'u-pan-she, Vols. 1-3, 1957; Vol. 4, Part A, 1959), 667 + 459 + 467 + 594 pp.

These four volumes constitute a comprehensive series which deal with (1) the ancient period; (2) Han; (3) Wei, Chin and the Southern and Northern Dynastues; and (4) Sui through the Ming, respectively, in a Marxist framework of interpretation. The text includes long quotations from the writers discussed, and this work is perhaps the most detailed available on its subject. Hou Wai-lu's Chung-kuo tsao-ch'i ch'i-meng ssu-hsiang (5.1.2) may be considered as the fifth and final volume of this series. An earlier version of this work, with the same title, was published in 1947 (Shanghai: Hsin-chih shu-tien) and reprinted in 1949 and 1950 (Peking: San-lien shu-tien). (20,000)

5.1.2 Hou Wai-lu 侯外廬, Chung-kuo tsao-ch'i ch'i-meng ssu-hsiang shih 中国早期啟蒙思想史 (A history of early "enlightenment" thought in China; Peking: Jen-min ch'u-pan-she, 1956), 5 + 689 pp.

The writers discussed in this volume are: (for the seventeenth century) Wang Fu-chih 王夫之, Huang Tsung-hsi 黃宗羲, Ku Yen-wu 顧炎武, Chu Chih-yü 朱之瑜, Fu Shan 傅山, Li Yung 李顒, and Yen Yuan 顏元; (for the eighteenth century) Tai Chen 戴震, Wang Chung 汪中, Chang Hsueh-ch'eng 章學誠, Chiao Hsün 焦循, and Juan Yuan 阮元; and (for the late eighteenth century and early nineteenth century) Kung Tzu-chen 龔自珍. The discussions are based on frequent and lengthy quotations. Hou asserts that, although the early Chinese "enlightenment" philosophers were different from their Russian counterparts of the nineteenth century (e.g., Cherneshevsky), yet both reflect the development of their respective societies in the direction of capitalism. On this premise the author goes on to describe and evaluate his Chinese exemplars. Since these men unfortunately have been little studied elsewhere, this work opens up an important fact of modern Chinese intellectual history. (12,000)

5.1.3 Chang Tai-nien 張岱年, Chung-kuo wei-wu chu-i ssu-hsiang chien-shih 中国唯物主义思想簡史 (A brief history of materialistic thought in China; Peking: Chung-kuo ch'ing-nien ch'u-pan-she, 1957), 131 pp. (60,000)

5.1.4 Lü Chen-yü 呂振羽, Chung-kuo cheng-chih ssu-hsiang shih 中国政治思想史 (A history of Chinese political thought; Peking: San-lien shu-tien, 1949), 20 + 507 pp.

A comprehensive Marxist survey of Chinese political thought, beginning with the Oracle Bones and ending with the eve of the Opium War, which discusses the "political philosophers" as products of their society (e.g., slave, feudal) and of their class origin (e.g., landlord, peasant). An earlier version of this book was published in Shanghai in 1937, and a new edition in 1955 by the publisher of the 1949 edition. (5,000)

5.1.5 Chung-kuo che-hsueh shih wen-t'i t'ao-lun chuan-chi 中国哲学史問題討論專輯 (A symposium on the question of the history of Chinese philosophy; Peking: K'o-hsueh ch'u-pan-she, 1957), 526 pp.

This volume of some fifty-five essays on Marxist philosophical questions is derived from a "Symposium on the history of Chinese philosophy" held by the Philosophy Department of Peking University in January 1957 (during the "Hundred Flowers" movement). Some of the essays (e.g., by Fung Yu-lan 馮友兰) suggest a preoccupation with the theories of idealism and a desire to reconcile them with the Marxist-dominated scene. This position is attacked in other essays. The question of how to treat the Chinese philosophic heritage is also discussed. (10,215)

5.1.6 Division of Teaching and Research on the History of Philosophy, Department of Philosophy, Chinese People's University, ed., Chung-kuo che-hsueh shih ts'an-k'ao tzu-liao 中国哲学史参考資料 (Reference materials on the history of Chinese philosophy; Peking: Chung-kuo jen-min ta-hsueh ch'u-pan-she, 1957), Vol. 1, 8 + 296 pp.

An elementary work intended to introduce the Chinese university student to original philosophical texts. It contains brief selections from "philosophical" sources, beginning with divinations in the Shang period and continuing through Confucius, Mencius, and the philosophical writings of the Warring States period and early Han. These materials are arranged mostly by topic--e.g., Mo-tzu on "universal love." For the selections from the Shang-shu and Shih-ching, vernacular translations are given. (23,692)

5.1.7 Hou Wai-lu 侯外廬, ed., Chung-kuo li-tai ta-t'ung li-hsiang 中国歷代大同理想 (The ta-t'ung ideal of China through the ages; Peking: K'o-hsueh ch'u-pan-she, 1959), 2 + 54 pp.

The four chapters of this small volume deal briefly with the idea of ta-t'ung ("universal harmony," roughly an equivalent of "utopia") in ancient and "medieval" China and in the modern era (ca. 1850-1911). Among the ancient proponents of ta-t'ung the authors include Mo-tzu and Lao-tzu. In the medieval period they point on the one hand to the popular egalitarian notions of peasant rebels like Li Tzu-ch'eng 李自成, and on the other hand to the "anarchic" and "atheist" society adovcated by the fourth-century Taoist radical Pao Ching-yen

鮑敬言 . The utopian ideals of the Taiping Kingdom and of K'ang Yu-wei are the examples used for the modern period. Throughout, the authors (under the chief editorship of Hou Wai-lu) proclaim their assumption that philosophers of the past were only capable of subjectively dreaming of a society free from class oppression, while Marxism-Leninism alone is scientifically formulated and can lead to a real utopia. (8,400)

5.1.8 History of Chinese Philosophy Section, Institute of Philosophical Research, Chinese Academy of Sciences, ed., Chung-kuo ta-t'ung ssu-hsiang tzu-liao 中国大同思想資料 (Source materials on Chinese "utopian" thought; Peking: Chung-hua shu-chü, 1959), 7 + 98 pp.

About forty short passages from writings spanning Confucius and Sun Yat-sen--including two novels--on ideal types of society. These selections are accompanied by explanatory notes and a summary in pai-hua. The editor's preface takes its cue from Lenin: Utopian ideals occur in a class society among the exploited and among certain intellectuals. These ideals are not attainable in the society that produced them, although they are extremely valuable as incentives to historical change. (7,500)

5.1.9 Yang T'ing-fu 楊廷福 , Ming-mo san ta ssu-hsiang-chia 明末三大思想家 (Three great thinkers of the end of the Ming dynasty [Ku Yen-wu, Huang Tsung-hsi, and Wang Fu-chih; Shanghai: Ssu-lien ch'u-pan-she, 1955), 158 pp., 3 plates. (?)

5.1.10 Chao Li-sheng 趙儷生 , Ku Yen-wu chuan lueh 顧炎武傳略 (A brief biography of Ku Yen-wu; Shanghai: Shang-hai jen-min ch'u-pan-she, 1955), 58 pp., 1 plate. (14,000)

5.1.11 Hsieh Kuo-chen 謝國楨, *Ku T'ing-lin hsueh-p'u* 顧亭林學譜 (A scholastic biography of Ku Yen-wu; Shanghai: Commercial Press, 1957), 216 pp.

This is a revised version of a work first published in 1930. The original literary style is retained, but in his revisions Hsieh claims to have benefited from recent studies of Ku Yen-wu (1613-82) by such Marxist-inspired historians as Hou Wai-lu (5.1.2) and Chao Li-sheng (5.1.10). The author's preface makes use of current Communist vocabulary. The text contains a biography of Ku; a summary of his views on learning, politics, history, literature, personal conduct, and authorship (with many quotations); a bibliography of Ku's works (with selections from prefaces and comments by other writers); and a section on Ku's "scholastic companions" and disciples. (4,000)

5.1.12 Hsieh Kuo-chen 謝國楨, *Huang Li-chou hsueh-p'u* 黃黎洲學譜 (A chronology of the scholarly life of Huang Tsung-hsi [1610-1695], rev. ed. (Shanghai: Commercial Press, 1956, first ed. 1932), 162 pp. (?)

5.1.13 Wang I 汪毅, *Wang Ch'uan-shan ti she-hui ssu-hsiang* 王船山的社会思想 (The social thought of Wang Fu-chih [1619-92]; Shanghai: Hsin-chih-shih ch'u-pan-she, 1956), 62 pp., 1 plate. (4,000)

5.1.14 Yang P'ei-chih 楊培之, *Yen Hsi-chai yü Li Shu-ku* 顏習齋与李恕谷 (Yen Yuan and Li Kung; Wuhan: Hu-pei jen-min ch'u-pan-she, 1956), 3 + 292 pp., 2 plates.

Evidently based on a close study of the writings of these two seventeenth-century thinkers, the core of this work discusses their political and philosophical theories, and their views of human nature and education. These chapters are preceded by biographical accounts of the two men, and are followed by a chapter on the influence of earlier philosophers (e.g., Wang Yang-ming 王陽明) and on the spread of the Yen-li school of thought. The author concludes that although this school sprang from a "petit-bourgeois landlord" background and had strong features of conservatism and idealism, it contained "active" and "progressive" elements and is consequently a rich jewel in the national treasure vault. Yen and Li are praised for their advocacy of economic measures to better the people's livelihood and for their opposition to the extreme idealism of Sung Neo-Confucianism. All this is in line with the effort to find traces of the beginnings of a bourgeoisie and bourgeois thought in the late Ming and early Ch'ing. (Note that the author denounces Hu Shih for applying the term "pragmatism" to Yen and Li.) (15,000)

5.1.15 Li Chih 李贄 , Ts'ang shu 藏書 (A book to hide; Peking: Chung-hua shu-chü, 1959), 2 vols., 35 + 1142 pp.; and Hsü ts'ang shu (Ts'ang shu continued), same publisher and date, 32 + 529 pp.

These are among the major works of Li Chih (hao, Cho-wu 卓吾 , 1527-1602), a heretical figure whose books were burned by the Ming government, banned by the Ch'ing, and who was put in prison where he committed suicide. These volumes contain biographical essays on about 1,000 historical figures from the Warring States period to the early Ming. While the historical data are derived from dynastic histories and other standard sources, the interpretation and classification are often peculiar to the author. In his preface, he comments on the personal and fluid qualities of the concept of right and wrong and challenges the straitjacketing acceptance of Confucian standards. By the same token he takes a contemptuous view of the Sung Neo-Confucianists. These and other anti-traditional views (e.g., on the position of women) have earned Li Cho-wu a eulogistic publisher's note and the current handsome reprinting in a punctuated text based on a Ming edition. (1,300)

5.2 INTELLECTUAL CURRENTS IN THE NINETEENTH CENTURY AND AFTER

Our remarks in the introduction to sec. 5.1 about the way in which intellectual history is studied and written in Communist China apply as well to the treatment of more recent periods. Again, the emphasis is upon the "progressive" currents of thought. The men who are singled out for attention are praised for their criticism of the "feudal order" and for their patriotism, but the esteem which they receive is always qualified. In the end their achievements must be measured by the standards of Marxist-Leninist-Maoist thought, and on these grounds they are invariably described as being subject to limitations arising from their historical milieu and class background. In this way the "reformist," "bourgeois," and "idealist" thinkers of the recent past are assimilated into the new historical picture that the Chinese Communist scholars are attempting to construct. Items 5.2.1 and 5.2.2 contain a large number of examples of the way in which recent intellectual figures are being treated in Chinese Communist historiography. They are followed by two collections of source materials which have been subjected to a similar ideological treatment. Items 5.2.7-11 are studies of the same kind devoted to individual persons. This section concludes with three items (5.2.13-15) which illustrate how thinkers who are on the other side of the fence, i.e. "anti-progressive," are treated. See also the works on the Reform Movement of 1898 and the May Fourth Movement in secs. 2.7 and 3.2 above, respectively.

5.2.1 Chung-kuo chin-tai ssu-hsiang shih lun-wen chi 中国近代思想史論文集 (A collection of essays on Chinese modern intellectual history; Shanghai: Shang-hai jen-min ch'u-pan-she, 1958), 210 pp.

This collection contains fifteen essays written in 1952 by half a dozen members of the philosophy department at Peita. Contributors include Fung Yu-lan 馮友兰 and Wang Wei-ch'eng 王維誠, and their subjects include Lin Tse-hsü 林则徐, Wei Yuan 魏源, Feng Kuei-fen 馮桂芬, Wang T'ao 王韜, Hsueh Fu-ch'eng 薛福成, Ma Chien-chung 馬建忠, Ho Ch'i 何啟, Hu Li-yuan 胡礼垣, Ch'en Chih 陳熾, Cheng Kuan-ying 鄭观應, K'ang Yu-wei 康有為, Liang Ch'i-ch'ao 梁啟超, Chang Ping-lin 章炳麟, Ts'ai Yuan-p'ei 蔡元培, Li Ta-chao 李大釗, and Ch'en Tu-hsiu 陳獨秀. There is also one article by Fung on Timothy Richard and others representing "imperialist colonialism." The ideas of each man are analyzed from the points of view of "anti-imperialism," "feudalism," "reformism," and the like. (6,500)

5.2.2 Chinese History Teaching and Research Section, Chinese People's University, ed., Chung-kuo chin-tai ssu-hsiang-chia yen-chiu lun-wen hsuan 中国近代思想家研究論文集 (Selected research articles on modern Chinese thinkers; Peking: San-lien shu-tien, 1957), 173 pp.

Liang Ch'i-ch'ao, T'an Ssu-t'ung 譚嗣同, Yen Fu 嚴復, and Sun Yat-sen are the four thinkers who are studied and judged in five articles by researchers connected with the Chinese People's University. These men are praised, in varying degrees, for their patriotism, their contributions to the decay of the "feudal" order and to the people's revolutionary stirrings, and other salutary functions. But when measured by the Communist value system, they--being subject to "historical, class, and cosmic limitations" (Lenin)--come through with less honorific labels: "reformist," "bourgeois," and "idealist." In short, these articles are primarily doctrinal exercises which also try to salvage something of the Chinese intellectual past. (37,000)

5.2.3 Shih Chün 石峻, ed., Chung-kuo chin-tai ssu-hsiang shih ts'an-k'ao tzu-liao chien-pien 中国近代思想史参考資料簡編 (A concise selection of reference materials on Chinese modern intellectual history; Peking: San-lien shu-tien, 1957), 17 + 1276 pp.

The readings selected for this volume are grouped under five headings: (1) "The rise of the ideologies of social reform among the landlord class and of revolution among the people after the intrusion of foreign capitalist forces" (represented by the writings of Kung Tzu-chen 龔自珍, Lin Tse-hsü, and Wei Yuan). (2) "The revolutionary ideology of the Taiping Kingdom during the rising tide of the peasant revolutionary movements" (e.g., Hung Hsiu-ch'üan 洪秀全, Yang Hsiu-ch'ing 楊秀清). (3) "Bourgeois reformist ideology during the formation of the semi-colonial, semi-feudal order" (e.g., K'ang Yu-wei, Liang Ch'i-ch'ao, Yen Fu, et al.). (4) "The development of revolutionary ideology among the bourgeoisie and petite bourgeoisie and its struggles with reformist ideology during the bourgeois revolutionary movements" (e.g., Chang Ping-lin, Ch'en T'ien-hua 陳天華, and Sun Yat-sen). (5) "The differentiation of bourgeois and petit bourgeois revolutionary ideology and the spread of Marxism in China during the May Fourth Movement" (e.g., Ts'ai Yuan-p'ei, Hu Shih, Ch'en Tu-hsiu, Lu Hsün 魯迅, and Li Ta-ch'ao. End notes (pp. 1263-76) contain biographical and other information on each of the authors represented. One may compare this collection with S. Y. Teng and J. K. Fairbank, China's Response to the West (1954), which covers much the same ground. (17,000)

5.2.4 History of Chinese Philosophy Section, Institute of Philosophical Research, Chinese Academy of Sciences, comp. and ed., Chung-kuo che-hsueh shih tzu-liao hsuan-chi--chin-tai chih pu 中国哲学史資料選輯--近代之部 (Selected materials on the history of Chinese philosophy--the modern period; Peking: Chung-hua shu-chü, 1959), 2 vols., 7 + 686 pp.

This work is an example of the monumental historical compilation projects, coupled with an attempt to pigeonhole everything into Marxist categories, which are being undertaken on the Chinese mainland today. These two volumes on the modern period are one part out of six in a series of materials on the history of "philosophy," the other five being devoted respectively to the Ch'in; Han; Six Dynasties, Sui, and Tang; Sung, Yuan, and Ming; and Ch'ing. (The selections of readings from the Ch'in and Han periods have been rendered into modern Chinese.) For the modern period the selections consist of about forty items by writers ranging in time from the Opium War to the May Fourth Movement, i.e., from Kung Tzu-chen and Wei Yuan through Hung Hsiu-ch'üan and Hung Jen-kan 洪仁玕, and ending with Sun Yat-sen and Lu Hsün. Only in the broadest sense can most of these writings be classified as "philosophy"; "Materials on certain trends (and not others) in modern Chinese thought" would be a far better title for the volumes. Each selection is accompanied by copious explanatory notes; biographical and critical (from a Marxist-Maoist viewpoint) data on the author; and an exegesis of the ideas contained therein. (4,000)

5.2.5 Shih Chün 石峻, Jen Chi-yü 任継愈, and Chu Po-k'un 朱伯崑, Chung-kuo chin-tai ssu-hsiang shih chiang-shou t'i-kang 中国近代思想史講授題綱 (A teaching outline for contemporary Chinese intellectual history; Peking: Jen-min ch'u-pan-she, 1955), 176 pp. (47,000)

5.2.6 Li Shih-yueh 李时岳, Chin-tai Chung-kuo fan yang-chiao yun-tung 近代中国反洋教运动 (The movement against foreign religions in modern China; Peking: San-lien shu-tien, 1958). (?)

5.2.7 Mai Jo-p'eng 麦若鵬, Huang Tsun-hsien chuan 黄遵憲傳 (A biography of Huang Tsun-hsien [1848-1905]; Shanghai: Ku-tien wen-hsueh ch'u-pan-she, 1957), 112 pp., illus.

> Poet, minor diplomatic official, and reformer associated with Liang Ch'i-ch'ao and T'an Ssu-t'ung, Huang is admired in mainland China for the emotional nationalism of his poetry. (?)

5.2.8 Ch'en Hsü-lu 陳旭麓, Tsou Jung yü Ch'en T'ien-hua ti ssu-hsiang 鄒容与陳天華的思想. (The ideology of Tsou Jung [1885-1905] and Ch'en T'ien-hua [1875-1905]; Shanghai: Shang-hai jen-min ch'u-pan-she, 1957), 60 pp.

> Ch'en T'ien-hua, a founder and the first secretary of the T'ung-meng Hui and an editor of Min-pao, committed suicide in Japan in 1905, supposedly as a protest against the expulsion from Japan of revolutionary Chinese students. Tsou was the author of Ko-ming chün (The revolutionary army), a violent attack on the Manchus which was published by the newspaper Su-pao in the Shanghai foreign settlement. See 3.1.7. (10,000)

5.2.9 Wang Shih 王栻, Yen Fu chuan 嚴復傳 (A biography of Yen Fu; Shanghai: Shang-hai jen-min ch'u-pan-she, 1957), 103 pp., illus. (?)

5.2.10 Ts'ai Shang-ssu 蔡尚思, Ts'ai Yuan-p'ei hsueh-shu ssu-hsiang chuan-chi 蔡元培学術思想記 (An account of the scholarship, thought, and life of Ts'ai Yuan-p'ei; Shanghai: T'ang-ti ch'u-pan-she, 1950), 20 + 464 pp., 2 plates. (2,000)

5.2.11 Ts'ai Yuan-p'ei hsuan-chi 蔡元培选集 (Selected writings of Ts'ai Yuan-p'ei; Peking: Chung-hua shu-chü, 1959), 7 + 336 pp.

Sixty-seven brief articles by Ts'ai Yuan-p'ei (1868-1940), educator, philosopher, and important figure in modern China's intellectual history. These are reprinted from newspapers and magazines and are arranged in chronological order, the earliest piece dating from 1902 and the latest from 1937. The selections cover a wide range of subjects, an indication of Ts'ai's many interests: educational philosophy, politics, aesthetics, and the new culture. The editor's preface gives a brief sketch of Ts'ai's career, pointing out that although as a philosopher he was an idealist and a "bourgeois intellectual," he was not (like Hu Shih and others) a "running dog of imperialism." Ts'ai in the current view was a "revolutionary democrat and a patriot," hence this new collection of his writings, containing many items not easily accessible elsewhere. (7,000)

5.2.12 Lo Ping-chih 羅炳之, Chung-kuo chin-tai chiao-yü-chia 中国近代教育家 (Educators of contemporary China; Wuhan: Hu-pei jen-min ch'u-pan-she, 1958), 4 + 208 pp. (9,800)

5.2.13 Hu Shih ssu-hsiang p'i-p'an 胡適思想批判 (Criticisms of Hu Shih's thought; Peking: San-lien shu-tien, 1955), 2 vols., 248 + 372 pp.

Over forty articles originally published in newspapers and periodicals during 1954-55 are reprinted in these two volumes, a few in somewhat revised form. The critics of Hu Shih include many important names among Chinese Communist writers (e.g., Kuo Mo-jo 郭沫若, Ai Ssu-ch'i 艾思奇, and Li Ta 李达); also professors who at one time or another had been influenced by Hu Shih (e.g., Lo Erh-kang 羅尔綱). Their criticisms deal not only with Hu's ideology and philosophy but also his scholarship, personality, role in the May Fourth Movement, relations with the imperialist West, and influence on Chinese intellectuals. (22,000)

5.2.14 Hu Feng wen-i ssu-hsiang p'i-p'an lun-wen hui chi 胡風文藝思想批判論文彙集 (Collection of critical essays on Hu Feng's literary ideas; Peking: Tso-chia ch'u-pan-she, 1955), 2 vols., 2 + 142 + 139 pp.

These two volumes represent two periods during which Hu Feng came under attack by orthodox Chinese Communist writers for his views on the role of literature. Vol. 1 consists of articles dating back to the Yenan period when Hu and such disciples as Shu Wu 舒蕪 were publishing their ideas in the magazine Hsi-wang 希望. Hence part of the criticism directed at Hu at that time was actually leveled at Shu and others, including Hu's mentor Lu Hsün. Vol. 2 contains articles written roughly between 1949 and 1953, some of the authors being Hu Feng's major opponents from the Yenan days, e.g., Mao Tun 茅盾 and Ho Ch'i-fang 何其芳. The last two articles in this volume are recantations by Shu Wu; they are included here presumably as a refutation of Hu. (100,000)

5.2.15 Liang Shu-ming ssu-hsiang p'i-p'an (lun-wen hui-pien) 梁漱溟思想批判論文彙編 (Critiques of Liang Shu-ming's ideology, a collection of essays; Peking: San-lien shu-tien, 1955), 2 vols., 173 + 237 pp.

A total of thirty-one articles, previously published in Chinese Communist newspapers and magazines during 1955, are reprinted here, some having undergone minor revisions by their authors. Various aspects of Liang's political, philosophical, and sociological theories are made the target of criticism by his fellow intellectuals--some of whom are also prominent in their own fields and have in their turn been criticized by others, though perhaps less systematically. One of these, the sociologist Wu Ching-ch'ao 吳景超, was already one of Liang's critics in the 1930's, when Wu was associated with the "renaissance liberal" group. The ideas for which Liang is attacked now are substantially the same as those that attracted criticisms in the 1930's--his views on Chinese society and culture, education, rural reconstruction, and related topics. The burden of the current criticism, however, is that he is anti-Marxist, anti-dialectical materialism, and therefore idealistic, reactionary, compradorish, and soon. (10,000)

5.3 PUBLISHING AND JOURNALISM

The six titles of this section are useful works, and in general are free of any particular ideological tinge. Items 5.3.3-5 are especially valuable for the lists of periodical publications which they contain.

5.3.1 Liu Kuo-chün 劉國鈞, Chung-kuo shu-shih chien-pien 中國書史簡編 (A concise history of books in China; Peking: Kao-teng chiao-yü ch'u-pan-she, 1958), vii + 159 pp., 20 plates. (5,500)

5.3.2 A Ying 阿英, Wan-Ch'ing wen-i pao k'an shu-lueh 晚清文藝报刊述略 (Brief descriptions of artistic and literary newspapers and periodicals in the late Ch'ing; Shanghai: Ku-tien wen-hsueh ch'u-pan-she, 1958), 156 pp., 8 plates. (4,000)

5.3.3 Chang Ching-lu 張靜廬, Chung-kuo chin-tai ch'u-pan shih-liao 中国近代出版史料 (Historical materials on publishing in modern China; Shanghai: [Vol. 1] Shang-tsa ch'u-pan-she, 1953; [Vol. 2] Ch'ün-lien ch'u-pan-she, 1954), 6 + 333 + 6 + 435 pp., many plates.

This collection of source materials on the Chinese publishing industry covers a period of over fifty years (1862-1918) from the establishment of the T'ung-wen Kuan 同文館 by the Ch'ing government. The materials, in the form of documents, essays, annotated bibliographies, and the like, relate to these topics: (1) official publishing and translation activities; (2) specific publications--newspapers, periodicals, and books; (3) technical aspects of printing, bookbinding, and the like; (4) laws governing publishing. The lists of publications should be of particular value to the historian; they include lists of late Ch'ing newspapers and periodicals, and of contemporary publications concerning such events as the Opium War, the Sino-Japanese War, and the Boxer Rebellion. Vol. 2 is designed as a supplement to Vol. 1; both cover the same period and contain the same types of material. A list of major publishing events from 1862 to 1918 is appended to Vol. 2. The many illustrations are for the most part pages reproduced from the publications under consideration. (3,000)

5.3.4 Chang Ching-lu 張靜廬, Chung-kuo hsien-tai ch'u-pan shih-liao

中国現代出版史料 (Historical materials on publishing in contemporary China; Peking: Chung-hua shu-chü, 1954-56), 3 vols., 41 + 468 pp., 43 + 527 pp., 48 + 534 pp., many plates.

A continuation of the previous item, Vol. 1 covers the period 1919-27; Vol. 2, 1927-37; and Vol. 3, 1937-49. The contents consist, for example, of (1) manifestos and pronouncements of "progressive" periodicals of the May Fourth period and after; (2) a list of Chinese translations of European, American, and Japanese literary works from the late Ch'ing until 1929; (3) materials on the republication of older texts; (4) bibliographical lists of revolutionary literature published in the 1920's and 1930's; (5) materials on "reactionary" laws concerning publishing; (6) revolutionary literature and its suppression; (7) publishing in the Communist-held areas; and (8) technological developments in printing and binding. (6,100)

5.3.5 Chang Ching-lu 張静廬, ed., Chung-kuo ch'u-pan shih-liao pu-pien 中国出版史料補編 (A supplementary collection of historical materials on publishing in China; Peking: Chung-hua shu-chü, 1957), 39 + 596 pp., illus.

This volume contains materials supplementary to the two collections on publishing for the "modern" and "contemporary" periods (5.3.3, 5.3.4). About seventy articles related to various aspects of the publishing industry are included. Most are reprinted from a variety of sources; others have not been published previously. Part I covers the "modern" period--1862-ca. 1919; Part II, the "contemporary" period, 1919-49. W.A.P. Martin's account of the T'ung-wen Kuan is included in translation in Part I; among the materials in Part II are lists of Chinese translations of the works of Marx, Lenin, and Stalin. The appendix (pp. 557-

596) includes lists of major events in the histories of two important publishing firms, the Commercial Press and Chung-hua Book Company; a chronology of the industry as a whole for 1918-49; and a short article on the Chinese paper industry. (5,000)

5.3.6 Ko Kung-chen 戈公振, Chung-kuo pao-hsueh shih 中国报学史 (A history of journalism in China; Peking: San-lien shu-tien, 1955), 11 + 378 pp., 1 plate.

A reprint of a pioneering account of the history of Chinese newspapers (first published in 1927), based on a 1935 Commercial Press edition. The author's own preface and some illustrations are omitted from this edition. There is a new foreword written by Ko's nephew, Ko Pao-ch'üan 戈宝权, which contains biographical data and also discusses the author's "progressive" tendencies, e.g., his interest in the USSR. (6,000)

5.4 LANGUAGE REFORM

Reform of the Chinese written language, in particular simplification of the characters and the much-discussed plans to introduce alphabetic writing, is one of the items on the agenda of the government of the People's Republic of China. The first three items in this section deal with some historical precedents for the current concern with language reform. Item 5.4.4 is a useful reference history of the Chinese language.

5.4.1 T'an Pi-an 譚彼岸, Wan-Ch'ing pai-hua wen yun-tung 晚清白話文運動 (The vernacular language movement in the late Ch'ing period; Wuhan: Hu-pei jen-min ch'u-pan-she, 1956), 44 pp.

Much of the credit for the vernacular movement in twentieth-century China is assigned to half a dozen or so early proponents, particularly to two scholars associated with the 1898 Reform,

Ch'iu T'ing-liang 裘廷梁 and Ch'en Jung-kun 陳榮袞. Their ideas, according to the author of this pamphlet, were flagrantly plagiarized by Hu Shih, and at least half of the text is devoted to substantiating this charge. The remaining portion discusses a number of vernacular publications prior to 1911. (26,000)

5.4.2 Ch'ing-mo wen-tzu kai-ko wen chi 清末文字改革文集 (A collection of writings on the language reform of the end of the Ch'ing; Peking: Wen-tzu kai-ko ch'u-pan-she, 1958), 4 + 144 pp. (1,900)

5.4.3 Ni Hai-shu 倪海曙, Ch'ing-mo Han-yü p'in-yin yun-tung pien-nien shih 清末汉語拼音运动編年史 (A chronological history of the Han language phonetic spelling movement at the end of the Ch'ing dynasty; Shanghai: Shang-hai jen-min ch'u-pan-she, 1959), 3 + 244 pp., illus., index.

The attempts made during the years 1891-1911 to introduce a system of writing Chinese with phonetic symbols, because of their connection with popular education, are now viewed as a high point of the "old democratic revolution." This volume begins with a brief chronological listing of the main events of the phonetic spelling movement. This is followed by a chart showing the several proposals made for reforming the writing system, with a few particulars about each. The remainder of the volume (pp. 13-236) discusses these proposals in greater detail--the authors, the proposed systems, and the outcome. There are illustrations of, among other things, some of the proposed symbols. The index (pp. 237-244) is arranged by roman letters as used in the "national language romanization" (kuo-yü lo-ma tzu 国語羅馬字). (2,500)

5.4.4 Wang Li 王力, Han-yü shih kao 漢語史稿 (A draft history of the Chinese language; Vol. 1, Peking: K'o-hsueh ch'u-pan-she, 1957), 5 + 209 pp. (Vol. 2, 1958, not seen). (16,160)

5.5 SCIENCE AND TECHNOLOGY

We have not attempted to cover comprehensively either the history of China's material culture or the scientific discoveries and technological innovations on which it was based. While this section includes a few examples of this genre of historical writing, the reader is referred to the historical sections of 5.5.1 for useful synopses of a larger number of works. If one general comment is to be made about the volumes listed here, it should perhaps be to note the strongly nationalistic flavor which characterizes them.

5.5.1 Chao Chi-sheng 赵繼生, comp., K'o-hsueh chi-shu ts'an-k'ao shu t'i-yao 科學技術参考書提要 (An annotated list of scientific and technical reference works; Peking: Commercial Press, 1958), 12 + 539 pp.

> The 1554 titles in this volume, many of which are translations from Russian or English, are arranged by subject. Works on the history of science and technology, including China's indigenous scientific tradition, are listed at the beginning of each section. (4,000)

5.5.2 Li Kuang-pi 李光璧 and Ch'ien Chün-hua 钱君曄, eds., Chung-kuo k'o-hsueh chi-shu fa-ming ho k'o-hsueh chi-shu jen-wu lun chi 中国科学技術發明和科学技術人物論集 (Collected essays on Chinese scientific and technological inventions and personages in these fields; Peking: San-lien shu-tien, 1955), 4 + 349 pp. (5,500)

5.5.3 Ch'ien Wei-chang 钱偉長 Wo-kuo li-shih shang ti k'o-hsueh fa-ming 我国历史上的科学发明 (A history of scientific discoveries in China; Peking: Chung-kuo ch'ing-nien ch'u-pan-she, 1953), 108 pp., illus.

A very simple account (presumably for readers below college level) of Chinese achievements in agriculture, engineering, mathematics, astronomy, and other scientific fields from ancient times on. Some reference works, both traditional and modern, are listed in a postscript; but no sources are indicated in the text. The author is a Western-trained scientist. (50,000)

5.5.4 Hung Kuang 洪光 and Huang T'ien-yu 黄天佑, Chung-kuo tsao chih fa-chan shih lueh 中国造紙发展史略 (A brief history of paper manufacture in China; Peking: Ch'ing-kung-yeh ch'u-pan-she, 1957), 52 pp. (1,400)

5.5.5 Chang Hsiu-min 張秀民, Chung-kuo yin-shua shu ti fa-ming chi ch'i ying-hsiang 中国印刷術的发明及其影響 (The invention of printing in China and its impact; Peking: Jen-min ch'u-pan-she, 1958), 6 + 208 pp., plates. (?)

5.5.6 Ch'en Tsun-wei 陳遵嬀, Ch'ing-ch'ao t'ien-wen i-ch'i chieh-shuo 清朝天文仪器解說 (An explanation of the astronomical instruments of the Ch'ing dynasty; Peking: All-China Association for the Popularization of Scientific Technology [Chung-hua ch'üan-kuo k'o-hsueh chi-shu p'u-chi hsieh-hui

], 1956), 59 pp., illus.

This small book was sponosred by the Astronomical Institute (T'ien-wen Kuan 天文館) of Peking, in connection with the opening to the public of the Old (Ming) Observatory in that city as part of a program of "patriotic education." The eight instruments housed in the observatory are described with the aid of numerous diagrams and photographs and against the background of a brief account of traditional Chinese astronomical instruments. The author's conclusion is that although the Ch'ing instruments were made under the supervision of Jesuit astronomers, they were still modeled on Chinese instruments of earlier dynasties. (20,000)

5.5.7 Wang Yung 王庸 , Chung-kuo ti-t'u shih kang 中国地圖史綱 (An outline of Chinese cartographic history; Peking: San-lien shu-tien, 1958), 4 + 112 pp., 10 plates. (5,200)

5.5.8 Chang Hsün 章巽 , Wo-kuo ku-tai ti hai-shang chiao-t'ung 我國古代的海上交通 (Seafaring in ancient China; Shanghai: Hsin chih-shih ch'u-pan-she, 1956), 42 pp., 7 maps. (6,000)

5.5.9 Chou Wei 周緯 , Chung-kuo ping-ch'i shih kao 中國兵器史稿 (A draft history of Chinese military weapons; Peking: San-lien shu-tien, 1957), 17 + 339 pp., 92 plates.

This is a substantial, detailed, and well-documented book, for which the author had been collecting materials for over thirty years, although he died before revising the manuscript, which was prepared for publication by the archaeologist Kuo Pao-chün 郭寶鈞 . The sources used include works on Asian weapons in Western languages as well as in Chinese and Japanese. There is no bibliography, but reference works are cited extensively in footnotes. There are illustrations of 854 weapons and other items of military equipment, either in photographs or in drawings in the ninety-two plates at the end of the book. In addition, sketches accompanying the text represent 128 weapons, making a total of 982 separate illustrations, many previously unpublished, drawn from weapons collections in museums, libraries, and private holdings. Pages 278-324 deal with the Ch'ing period. (3,000)

5.6 LITERATURE AND ART

This section, like the last, is only a sampling of the many works now available on these subjects. Item 5.6.1 is a particularly noteworthy survey. Items 5.5.5-8 represent the increasingly stressed field of contemporary or "new" literature, which the Ministry of Education in 1950 established as a required course for all university students of Chinese language and literature. The contents of such a course were specified by the Ministry as "Lectures on the course of development of China's new literature from the May Fourth Movement to the present, using the new standpoint and new method, emphasizing the struggles in belletristic ideology in each period and the way in which these developed, as well as a critical account of the famous authors and works of essays, poems, plays, and novels."

5.6.1 Cheng Chen-to 鄭振鐸, Ch'a-t'u-pen Chung-kuo wen-hsueh shih 插畫本中国文学史 (An illustrated Chinese literary history; Peking: Tso-chia ch'u-pan-she, 1957), 4 vols., 19 + 1024 pp., many plates.

A revised version of the 1932 edition, one of the earliest and still among the most valuable scholarly treatments of this subject. (2,600)

5.6.2 Lin Keng 林庚 , Chung-kuo wen-hsueh chien-shih 中国文学简史 (A brief history of Chinese literature; Shanghai: Ku-tien wen-hsueh ch'u-pan-she, 1957), 2 vols., Vol. 1, 3 + 386 pp. (Vol. 2 not seen). (20,000)

5.6.3 Chung-kuo wen-hsueh shih 中国文学史 (Chinese literary history [written by the literature concentrators of the class of 1955, Department of Chinese, Peking University]; Peking: Jen-min wen-hsueh ch'u-pan-she, 1958), 2 vols., 5 + 401 + 6 + 700 pp.

This work was undertaken by some thirty students at Peking University during their summer vacation in 1958, in response to a call from the CCP Committee of the University to participate in the "Great Leap Forward." The writing of the manuscript was completed in twenty-four days. The writers are concerned, on the one hand, with extolling the great literary tradition of China and, on the other, with literature primarily as a reflection of society. Proportionally, folk literature and works with elements of social significance receive much more attention than those which do not lend themselves as readily to discussion in Marxist or nationalistic terms. (35,000)

5.6.4 A Ying 阿英 , Wan-Ch'ing hsiao-shuo shih 晚清小說史 (A history of the novel in the late Ch'ing; Peking: Tso-chia ch'u-pan-she, 1955, first pub. 1937), 190 pp. (?)

5.6.5 Wang Hsiao-chuan 王曉傳, comp., Yuan Ming Ch'ing san-tai chin hui hsiao-shuo hsi-ch'ü shih-liao 元明清三代禁毀小說戲曲史料 (Historical materials on the proscription and destruction of novels and dramas in the Yuan, Ming and Ch'ing periods; Peking: Tso-chia ch'u-pan-she, 1958), 52 + 360 pp. (2,600)

5.6.6 Wang Yao 王瑤, Chung-kuo hsin wen-hsueh shih kao 中国新文学史稿 (A draft history of China's new literature; Shanghai: Hsin-wen-i ch'u-pan-she, 1953, first pub. 1951), 2 vols., 10 + 310 + 5 + 543 pp.

A detailed and extremely useful Communist account of literary history from 1919 to 1952. Wang's theme is the ever increasing role of literature in political struggle. His periodization is based in part on Mao Tse-tung's treatment of "the United Front in the cultural revolution" as outlined in On New Democracy. The period 1937-52 is covered in greater detail than is the period 1919-37. (14,000)

5.6.7 Ting I 丁易, Chung-kuo hsien-tai wen-hsueh shih lueh 中国現代文学史略 (A brief history of contemporary Chinese literature; Peking: Tso-chia ch'u-pan-she, 1955), 10 + 456 pp. (92,000)

5.6.8 Liu Shou-sung 刘綬松, Chung-kuo hsin wen-hsueh shih ch'u kao 中国新文学史,初稿 (A history of China's new literature, first draft; Peking: Tso-chia ch'u-pan-she, 1956), 2 vols., Vol. 2, 342 pp. (Vol. 1 not seen). (20,000)

5.6.9 Hu Wen-k'ai 胡文楷, Li-tai fu-nü chu-tso k'ao 歷代婦女著作考 (A study of works by female authors; Shanghai: Commercial Press, 1957), 902 pp. (4,000)

5.6.10 Shih Ching 史靖, Wen I-to 聞一多 (Wuhan: Hu-pei jen-min ch'u-pan-she, 1958), 116 pp.

A biography of one of the leading poets of modern China (1899-1946), who was also a sharp critic of the Kuomintang. (12,000)

5.6.11 Li Yü 李浴, Chung-kuo mei-shu shih kang 中国美術史綱 (An outline of Chinese art history; Peking: Jen-min mei-shu ch'u-pan-she, 1957), 340 pp., 76 plates. (7,000)

6. REFERENCE WORKS

Included in Section 6 are a number of "tools" which should facilitate the use of the substantive volumes analyzed in the preceding sections of this bibliography. Others, especially the bibliographies and indices in sec. 6.1, will, we hope, help the reader to go beyond the confines of our selections and assemble a comprehensive list of books and articles published in Communist China which deal with the topic in which he is particularly interested. Even works of reference, however, are not free of ideology, which governs what books shall be published and often influences their content. Thus, for example, the publisher of a reprint of an index to the important Ming biographical collections asserts that "we are too busy today with more pressing and important tasks" to revise the index in question. "Emphasizing the present and de-emphasizing the past" gives a relatively low priority to the study of the Ming Dynasty. Or, as another example, the editor of an extremely valuable check list of local gazetteers refers to the rare editions of _ti-fang chih_ in the Library of Congress as having been "pilfered" from China.

6.1 BIBLIOGRAPHIES AND INDICES

Many of the bibliographical tools published in the People's Republic of China are not generally available in Europe or America. Copies of the two national bibliographies and the major periodical index (6.1.1-3), for example, are extremely difficult to obtain. While several mainland Chinese libraries have published catalogues of one kind or another, we have seen only 6.1.4 .

The remainder of this section lists some bibliographical reference works which are of particular value for the study of Chinese history. Note especially 6.1.6 and 6.1.7. The publication of two additional volumes which continue 6.1.7 from 1937 down to the present has been promised, but we have not seen this publication if it has appeared. Items 6.1.8-12 are each probably the most comprehensive publications of their kind. Item 6.1.16 is the fullest bibliography available on the Taiping Rebellion. On the May Fourth Movement, see 3.2.7. For works on the history of science in China, see 5.5.1.

6.1.1 Ch'üan-kuo tsung shu-mu 全国總書目 (National bibliography of China; Peking: Hsin-hua shu-tien).

An annual listing of nearly all books, maps, and pictures published in Communist China, arranged by subject. The title, author and/or translator, price, publisher, date, and number of copies issued of each publication are given. A composite volume for 1949-54 was published in 1955; thereafter annual volumes have appeared at irregular intervals.

6.1.2 Ch'üan-kuo hsin shu-mu 全国新書目 (A listing of new books in China; Peking: Ministry of Culture, Library of the Publishing Industry Supervisory Bureau).

A periodical listing of new books published in Communist China, giving for each the title, author, publisher, size, number of pages, number of characters, number of copies issued, date, and price. One fascicle appeared in 1950, four in 1951, two in 1952, and one in 1953. Since the beginning of 1954, this bibliography has been issued monthly.

6.1.3 Ch'üan-kuo chu-yao ch'i-k'an tzu-liao so-yin 全国主要期刊資料索引 (Index to articles in the major periodicals of China; Shanghai: Newspaper and Periodical Library of the City of Shanghai).

This index to magazine and newspaper articles has been published bi-monthly since 1955. The contents are arranged under subject headings; and for each entry the title, author, translator, magazine or newspaper in which it appeared, date, volume, issue, and page numbers are given. Approximately 5,000-6,000 articles are indexed in each issue.

6.1.4 Peking University Library, comp., Pei-ching ta-hsueh t'u-shu-kuan Chung-wen chiu ch'i-k'an mu-lu 北京大学圖書館中文旧期刊目錄 (A list of Peking University Library's holdings in old [pre-1949] Chinese periodicals), 3 vols., Vol. 1, 1956; Vol. 2, 1957 (Vol. 3 not seen). (?)

6.1.5 Chang Shun-hui 張舜徽, Chung-kuo li-shih yao chi chieh-shao 中国歷史要籍介紹 (An introduction to the major works of Chinese history; Wuhan: Hu-pei jen-min ch'u-pan-she, 1956), 14 + 229 pp.

Written as a textbook for second-year students in the Department of History of Central China Normal College, where the author teaches, this book gives an overall view of the basic historical literature of China from the Shang dynasty through the Ch'ing. Following an introduction to the archaeological and literary sources for ancient history, the major traditional historical works (and a very small number of modern studies) are discussed under these categories: general histories (t'ung-shih 通史); dynastic histories (tuan-tai shih 斷代史); political histories (the tzu-chih t'ung-chien 資治通鑑 and chi-shih pen-mo 紀事本末 genres); cultural and institutional histories (the t'ung-tien 通典, wen-hsien t'ung-k'ao 文獻通考, and hui-yao 會要 genres); and local histories (fang-chih 方志). Some biographical data on the important historians are given in connection with their works. In addition, there is a short chapter on Chinese geography, and a long one on historiography, with Liu Chih-chi's 劉知幾 Shih-t'ung 史通 and Chang Hsueh-ch'eng's Wen-shih t'ung-i 文史通義 as the principal examples. The final chapter lists by title, with critical comments, the "tools" for the study of Chinese history which the author recommends (e.g., dictionaries, indices, catalogues, specific editions of the classics, pi-chi, encyclopediae, and a basic historical library). This is a useful book for beginners, even though one might not always agree with the author's evaluation of specific works. The Marxist "classics" are not included because, the author states, he assumes that all students of history will read them. (10,000)

6.1.6 Wang Chih-chiu 王芝九 and Sung Kuo-chu 宋國柱, eds., Chung-hsueh li-shih chiao-shih shou-ts'e 中学历史教師手册 (Handbook for middle school history teachers; Shanghai: Shang-hai chiao-yü ch'u-pan-she, 1958), 3 + 408 pp.

Part I (pp. 1-81) of this volume provides ideological guidance to the middle school teacher (e.g., to carry on education in historical materials on the basis of a firm grasp of Marxism-Leninism-Maoism, and to conduct political education in patriotism, internationalism, community morality, and the dignity of labor), and also offers detailed suggestions about teaching techniques and devices in and out of the classroom. Part II contains a "Chronological table of major historical events in China and foreign countries, 3500 B.C.-A.D. 1957" (pp. 83-236), the accuracy and usefulness of which is at times open to question. This followed by a list of Chinese dynasties and reign periods (pp. 237-258), and a series of lists of dynasties and rulers for other countries (pp. 259-274). Pages 275-301 contain a list of historical works published in the People's Republic of China between October 1949 and December 1957--a very useful bibliography of about 625 items, including works on foreign history which are preponderantly translations from the Russian. The concluding section (pp. 302-408) is a bibliography, arranged by subject, of articles on Chinese and foreign history published during the same period--about 2500 items. These two bibliographies should be of considerable interest to researchers. (11,000)

6.1.7 First and Second Offices of the Institute of Historical Research, Chinese Academy of Sciences and the Department of History, Peking University, comp., Chung-kuo shih-hsueh lun-wen so-yin 中国史学論文索引 (An index to articles on Chinese history; Peking: K'o-hsueh ch'u-pan-she, 1957), 2 vols., 6 + 421 + 8 + 676 + 115 pp.

This impressive index is a classified list of more than 30,000 articles dealing with Chinese history published in some 1,300 Chinese periodicals between 1900 and July 1937. Based in part on Kuo-hsueh lun-wen so-yin 国学論文索引 (Index to articles on Chinese studies) compiled by the Peking Library and other earlier indices, the present work has been augmented by a search of the periodical holdings in the collections of the several libraries in Peking. Vol. 1 contains a list of the periodicals indexed (with publisher, inaugural date, and issues available). The articles are grouped in four major sections with numerous subdivisions: (1) Chinese history, arranged by genre and by chronological sequence; (2) biography; (3) archaeology; (4) bibliography and reference. The major divisions in Vol. 2

(also with numerous subdivisions) are (1) intellectual history, (2) social history and the history of political institutions, (3) economic history, (4) cultural and educational history, (5) religions, (6) language, (7) literature, (8) the fine arts, (9) historical geography, (10) natural science, (11) agriculture, (12) medicine, and (13) engineering and technology. Articles translated from foreign languages are also included. A list of English names, together with Chinese transliterations, is appended. There is an index of subjects, personal and place names, special terms, and the like, arranged by number of strokes (111 pp.), but the absence of an index of authors is a major defect. (6,685)

6.1.8 Chu Shih-chia 朱士嘉, ed., Chung-kuo ti-fang-chih tsung-lu tseng-ting pen 中国地方志綜錄增訂本 (A comprehensive list of Chinese local gazetteers), rev. and enlarged ed. (Shanghai: Commercial Press, 1958, first ed. 1935), 7 + 318 + 105 pp.

This revised and enlarged edition of Chu's important guide to local gazetteers incorporates over 700 items not found in the first edition nor in the "Supplement" (pu-pien 补編, compiled by the editor and published in Shih-hsueh nien-pao 史学年报 ["Historical Annual"], Vol. 2, No. 5, 1938). It corrects errors in the first edition, and takes into account the changes in the repositories that have occurred since 1938. Items in the earlier edition that are ascertained to be lost are now deleted. Altogether a total of 7,413 titles (109,143 chüan) located on the mainland are listed in this volume. In addition, 232 titles (3,487 chüan) of "rare gazetteers pilfered and shipped to Taiwan," and about 800 rare items (out of a total of about 4,000) "pilfered" by the U.S. Library of Congress are listed in an appendix. There is a 105-page index of titles and authors arranged by number of strokes. (2,200)

6.1.9 Shanghai Library, comp., Chung-kuo ts'ung-shu tsung-lu 中国丛書綜錄 (A comprehensive list of the ts'ung-shu of China; Peking: Chung-hua shu-chü, Vol. 1, 1959), 3 vols., 4 + 1186 pp.

A monumental list covering 2797 classical ts'ung-shu found in 41 libraries in China (including the Peking National Library). Vol. 1 lists: (A) general collections (hui-pien 彙編), arranged chronologically, and (B) classified collections (lei-pien 類編), arranged by subject. These listings are followed by a long table (pp. 957-1133) which indicate the title, compiler, edition, and location (library) of each of the 2797 ts'ung-shu. There is an index of titles arranged by the "four corner" method, and another by number of strokes. Vols. 2 and 3, not yet available at the time of this writing, will be devoted to a listing of the individual works in the ts'ung-shu and appropriate indices. (1,600)

6.1.10 Teng Yen-lin 鄧衍林, comp., Chung-kuo pien-chiang t'u chi lu 中国边疆圖籍錄 (A list of maps and books on China's border areas; Shanghai: Commercial Press, 1958), 4 + 329 + 64 + 37 + 4 pp.

The manuscript of this book was completed some twenty years ago and is now published for the first time. The "border areas" are defined by the compiler as including China's Inner Asian frontiers and the areas of the minority races. Nearly 10,000 titles relating to both the history and geography of these areas are listed. The "barbarian dynasties" (e.g., Liao, Chin, Yuan) are given separate treatment. The arrangement of the material is by subject or by geographic area, and chronologically within these major divisions. Each entry includes the title, number of volumes, author, and edition. For rare editions, the location is also given (e.g., books at the Peking National Library are marked with an asterisk). Books not now extant are so indicated, together with the source of the citation. Only principal gazetteers are included; for fu and hsien gazetteers, the reader is referred to Chu Shih-chia's listing (6.1.8). An index by author and another by title are arranged according to the "four-corner" method. A guide to the four-corner numbers is also provided, arranged by number of strokes. (2,200)

6.1.11 Yang Tien-hsün 楊殿珣, comp., Shih-k'o t'i-pa so-yin 石刻題跋索引 (An index to stone-carved colophons; Shanghai: Commercial Press, 1957, first pub. 1941), 10 + 807 pp. (1,300)

6.1.12 Legal History Research Office, Bureau of Legal Affairs, State Council, comp., Chung-kuo fa-chih shih ts'an-k'ao shu-mu chien chieh 中国法制史参考書目简介 (A brief introduction to reference works on Chinese legal history; Peking: Fa-lü ch'u-pan-she, 1957), 228 pp.

A descriptive bibliography consisting of 932 titles (10,607 ts'e) in the possession of the Bureau of Legal Affairs. These books are arranged under ten categories: (1) works of the Legalists, (2) historical materials on the legal system, (3) codes and ordinances, (4) precedents and regulations, (5) the hui-yao and hui-tien of the several dynasties, (6) investigations and evidence, (7) trials and judgments, (8) prison administration, (9) collections of official correspondence, and (10) "others." The content of each item is briefly described. A number of works of the Kuomintang era, up to about 1945, are included. (2,500)

6.1.13 Ch'en Yuan 陳垣, Chung-kuo fo-chiao shih chi kai-lun 中国佛教史籍概論 (A brief introduction to works on the history of Chinese Buddhism; Peking: K'o-hsueh ch'u-pan-she, 1955), 5 + 149 pp.

The manuscript of this book, developed from Professor Ch'en's university lectures, was completed in 1942 but not published until 1955. The work consists of a systematic description of the basic sources for Buddhist history between the Six Dynasties and the Ch'ing, with particular attention to their value for students of Chinese history in general. These sources are introduced chronologically, with data on each item as to the title(s), compiler(s), edition(s), contents, and other aspects relevant to historical research. Erroneous statements about these works in the Ssu-k'u t'i-yao 四庫提要 are also pointed out and discussed. (2,835)

6.1.14 First and Second Offices, Institute of Historical Research, Chinese Academy of Sciences, ed., Shih-chi yen-chiu ti tzu-liao ho lun-wen so-yin 史記研究的資料和論文索引 (An index of source materials and articles for research on Shih-chi; Peking: K'o-hsueh ch'u-pan-she, 1957), 2 + 68 pp., 9 plates.

This useful and comprehensive bibliography contains lists of: (1) all the extant editions of Shih-chi (SC); (2) other indices to SC; (3) general introductions to SC; (4) studies of SC in toto (e.g., textual, authentification, commentaries); (5) studies of parts of SC (e.g., pen-chi 本紀, nien-piao 年表); (6) biographical and critical works on Ssu-ma Ch'ien, including portraits, accounts of his tomb, and family genealogy; (7) lost works and unpublished manuscripts on SC; (8) "non-specialized" works that touch on SC; (9) items in the pi-chi of the Sung, Yuan, and Ming periods which concern SC; (10) works on SC in foreign languages (predominantly Japanese essays). There are photographs of the tomb of Ssu-ma Ch'ien in Shensi and of pages of some rare editions of Shih-chi. (11,270)

6.1.15 Ho Tz'u-chün 賀次君, Shih-chi shu-lu 史記書錄 (A list of [sixty-four] editions of the Shih-chi [with commentaries]; Peking: Commercial Press, 1958), 9 + 234 pp. (1,100)

6.1.16 Chang Hsiu-min 張秀民, Wang Hui-an 王会庵, comps., T'ai-p'ing t'ien-kuo tzu-liao mu-lu 太平天国資料目錄 (A bibliography on the Taiping Kingdom; Shanghai: Shang-hai jen-min ch'u-pan-she, 1957), 4 + 224 pp.

This is published as a supplement to the six-volume T'ai-p'ing t'ien-kuo (2.5.5). Four types of materials are included: (1) documents of the Taiping Kingdom (ca. 400 items), (2) records from the Ch'ing side (600), (3) modern works in Chinese (370), and (4) works by foreign authors (70). About forty items on the T'ien-ti Hui 天地會 are appended. The compilers have supplied information on the edition(s) and location for most entries, and have also compiled a list of lost Taiping works and a list of forgeries. Among the recent Chinese works are articles and historical novels, in addition to surveys and monographic studies. The foreign works section, which includes a list of Chinese translations, is quite full with respect to contemporary English-language books, but is weak on Japanese studies. (30,000)

6.1.17 Ting Ching-t'ang 丁景唐 , Ch'ü Ch'iu-pai chu i hsi-nien mu-lu 瞿秋白著訳系年目錄 (A chronological list of the writings and translations of Ch'ü Ch'iu-pai; Shanghai: Shang-hai jen-min ch'u-pan-she, 1959), 156 pp., plates.

Ch'ü (1899-1935) succeeded Ch'en Tu-hsiu in August 1927 as secretary-general of the CCP, and in turn was reprimanded by the Sixth National Congress (July-September 1928) for "putschism" and removed from the leadership. He is honored in Communist China today more for his martyrdom and his literary writings than for his leadership of the Party. (?)

6.2 CHRONOLOGIES AND CONCORDANCES OF DATES

The works included in this section are either lists of events in Chinese history or concordances designed to aid in the conversion of dates in the Chinese lunar calendar into corresponding dates in the Western solar calendar. Items 6.2.3 and 6.2.4 are concerned explicitly with modern history. See also 6.1.6.

6.2.1 Jung Meng-yuan 荣孟源 , comp., Chung-kuo li-shih chi-nien 中国历史紀年 (A chronology of Chinese history; Peking: San-lien shu-tien, 1956), 3 + 274 pp.

Part I, "A list of reigns in the successive dynasties," 206 B.C. to 1949, gives the personal name, temple name and other titles, and the reign period for each of the emperors from the Han dynasty down to the Republic. Discrepancies in the sources are noted. Of especial interest is the inclusion of similar data for the rebel leaders and imperial pretenders at the end of each of the dynastic sections. Part II, "A chronology of the successive dynasties," contains fifteen tables which give the Western date, the Chinese cyclical designation, and the reign period for every year between 841 B.C. and 1949. The treatment of the Chou states, the "Six Dynasties" and the "Five Dynasties" is very full. Part III is an index of all the reign titles listed in Part I, arranged by number of strokes. (10,000)

6.2.2 Wan Kuo-ting 萬國鼎, comp., revised by Wan Ssu-nien 萬斯年 and Ch'en Meng-chia 陳夢家, Chung-kuo li-shih chi-nien piao 中国歷史紀年表 (Chronological tables of Chinese history; Shanghai: Commercial Press, 1956), 166 pp.

This work is based on a pre-war compilation by Wan Kuo-ting, Chung-Hsi tui-chao li-tai chi-nien t'u-piao 中西对照歷代紀年圖表 (Tables of Chinese-Western dates for the successive dynasties), which is now out of print. In the present revised and augmented version, Part I consists of a single long table giving the Chinese dynasty, reign, and cyclical characters and the corresponding year of the Christian calendar for each year in the period 849 B.C. to A.D. 1949. Part II includes a series of shorter tables giving the chronology of each of the Chinese dynasties (including detailed analyses of the chronology of the many states of the Chou period and of the temple names, personal names, and reign periods of each emperor), and a Japanese-Chinese comparative chronology. There is an index of states, imperial temple names (miao-hao 廟号), reign titles (nien-hao 年号), and personal names, arranged by the number of strokes. (4,000)

6.2.3 Department of Modern Chinese History, Kirin Normal University, comp., Chung-kuo chin-tai shih shih chi 中国近代史事記

(A chronological list of events in Chinese modern history; Shanghai: Shang-hai jen-min ch'u-pan-she, 1959), 442 pp.

Brief entries (usually a sentence or two) are made for notable dates during the years 1839-1919, almost unamimously agreed upon by mainland Chinese and Soviet historians as the "modern period" (chin-tai) of Chinese history. The Chinese lunar date is given in parentheses after each Western date. The chronology emphasizes domestic politics and foreign relations; economic and cultural happenings are scantily covered. A preliminary seven-page background list for the years 1514-1838 deals entirely with foreign contact and opium. (9,000)

6.2.4 Jung Meng-yuan 荣孟源, comp., Chung-kuo chin-tai shih li piao 中国近代史曆表 (Calendrical tables for Chinese modern history; Peking: San-lien shu-tien, 1953), 127 pp.

This volume, based on Ch'en Yuan's 陳垣 Erh-shih shih shu-jun piao 二十史朔閏表 (1926) consists of 120 tables, one for each year, from 1830 to 1950. Each table gives the days and the months by both the Western and the Chinese calendar (including the sexagenary cycles and the twenty-four solar "festivals"), so that, having the date by one reckoning, one can easily find the corresponding date by the other calendar. For example, the Chinese date for the Treaty of Nanking, "the twenty-second year of Tao-kuang, seventh moon, twenty-fourth day," after one finds the appropriate column and row, is converted to August 29, 1842. The appendix gives the calendar of the Taiping Kingdom, a table for computing the day of the week for any date from A.D. 1 to 5000 and a table of rhymes (used in telegrams and newspapers for days of the month). (3,000)

6.2.5 Ch'en Meng-chia 陳夢家, <u>Liu-kuo chi-nien</u> 六國紀年 (A chronology of the Six States of the Warring States period; Shanghai: Hsueh-hsi sheng-huo ch'u-pan-she, 1955), 146 pp. (6,000)

6.2.6 Ch'i Ssu-ho 齊思和 et al., <u>Chung wai li-shih nien piao</u> 中外历史年表 (Chronological tables of Chinese and Western history, 500 B.C.-A.D. 1918; Peking: San-lien shu-tien, 1958), iv + 883 pp. (?)

6.3 BIOGRAPHICAL AND GEOGRAPHICAL AIDS

Along with the production of new reference books, publishers in mainland China have been reprinting older standard works, such as 6.3.1-3. Item 6.3.4, on the other hand, is an extremely important new biographical guide for the Ch'ing dynasty. Items 6.3.7 and 6.3.8 are useful aids for identifying foreign place names that have been transliterated into Chinese characters.

6.3.1 Ting Chuan-ching 丁傳靖, comp., <u>Sung-jen i-shih hui-pien</u> 宋人軼事彙編 (A collection of biographical notes on persons of the Sung dynasty; Peking: Commercial Press, 1958, first pub. 1935), 2 vols., 16 + 1063 + 24 pp. (1,600)

6.3.2 <u>Erh-shih-wu shih jen-ming so-yin</u> 二十五史人名索引 (Index to personal names in the Twenty-five Standard Histories; Peking: Chung-hua shu-chü, 1956), 518 pp.

A reprint of the index to personal names originally published in 1935 to accompany the K'ai-ming 開明 edition of the standard histories. The present edition corrects typographical errors in the original. The "twenty-fifth" history is of course the Hsin Yuan-shih 新元史 (New Yuan history). (3,000)

6.3.3 Pa-shih-chiu chung Ming-tai chuan-chi tsung-ho yin-te 八十九種明代傳記綜合引得 (Combined indices to eighty-nine Ming dynasty biographical collections; Peking: Chung-hua shu-chü, 1959), 3 vols., 138 + 326 + 281 pp.

A photographic reprint of No. 24 in the Harvard-Yenching Institute Sinological Index Series (1935), without acknowledgment and without any indication that the original compiler was T'ien Chi-tsung 田繼綜. The publisher's note remarks that "we are too busy today with more pressing and more important tasks" to undertake the revision that these volumes need. (900)

6.3.4 Ch'en Nai-ch'ien 陳乃乾, Ch'ing-tai pei-chuan wen t'ung-chien 清代碑傳文通檢 (A finding list of epitaphs and biographies of the Ch'ing period; Peking: Chung-hua shu-chü, 1959), 5 + 410 pp.

This is a finding list for biographical information about more than 12,000 persons of the Ch'ing period, including individuals born in the Ming but still alive in 1644 and individuals who lived beyond 1911. The names of these 12,000 are arranged in order by number of strokes. After each name appear the tzu and hao, the place of birth, the dates of birth and death, and a reference to a biographical source or sources. These sources are biographies, epitaphs, prefaces, and other types of essays yielding biographical information which the compiler has extracted from a perusal of over 1,000 works of the Ch'ing period. An appendix (pp. 380-410) contains: (1) a list of variant names for ca. 700 persons, (2) a list of persons whose birth and/or death dates are given variously in different sources, and (3) a list of 1153 works by Ch'ing authors which Mr. Ch'en has personally seen (including 128 that he has not used for the present volume). (1,300)

6.3.5 Ch'en Yuan 陳垣, Shih hui chü-li 史諱舉例 (Examples of historical "name-taboos"; Peking: K'o-hsueh ch'u-pan-she, 1958), 9 + 175 pp.

The custom of avoiding the use of any characters which formed part of an emperor's name has led to all sorts of ambiguities and errors in Chinese historiography. The present work is intended to clarify some of these problems and to aid in dating historical records and in authenticating texts. In addition to discussing the types of name-taboos and the many problems that these created, the author lists the specific characters which were tabooed or altered in successive dynasties. This work was first published in 1928. The current edition is revised and contains more precise references. The bibliography lists over 100 works. (5,385)

6.3.6 Wang Shu-shih 王潄石, Chung-kuo li-shih ti-t'u 中国历史地圖 (Historical maps of China; Shanghai: Hsin-ya shu-tien, 1953), 6 maps. (?)

6.3.7 Chung-O-Ying shih-chieh chiao-t'ung ti-ming tz'u-tien 中俄英世界交通地名辞典 (A Chinese-Russian-English dictionary of geographical names used in world communications; Peking: Jen-min chiao-t'ung ch'u-pan-she, 1955), 249 pp. (pocketbook size). (2,600)

6.3.8 Han-O-Ying tui-chao ch'ang-yung wai-kuo ti-ming ts'an-k'ao tzu-liao 漢俄英对照常用外国地名参攷資料 (A Chinese-Russian-English dictionary of common foreign geographical names; Peking: Ti-t'u ch'u-pan-she, 1959), 167 pp.

Part I contains the Chinese versions for common foreign place names, followed by the Russian and then the English equivalents, the location (longitude and latitude), and in some cases an alternative Chinese equivalent. The arrangement is by the alphabetical order used in the kuo-yü romanization. Part II is from Russian to Chinese, and Part III from English to Chinese. Appended are six tables prepared by the translation section of the Hsinhua News Agency for converting the pronunciation of letters of the Russian, English, German, French, Italian, Spanish, and Yugoslav alphabets into Chinese characters and phonetic symbols. The editor's prefatory note explains some of the principles followed in the proposed standardized Chinese renditions. (3,000)

6.4 HISTORIOGRAPHY AND HISTORICAL PERIODICALS

The first three items in this section are examples of a large number of "theoretical" works which have appeared during the last decade in Communist China, and which are designed to instruct the student in the Marxist-Leninist approach to history. We have not attempted to cover such materials in this bibliography. Even reference books can be ideological, as indicated for example in the cases of 6.4.5 and 6.4.6. A number of important works of Ch'ing historical scholarship have recently been reprinted (for example, 6.4.10-14), partly because the scholarship of these works is still useful, but also for nationalistic reasons. Nationalism and the effort to incorporate the "best" of the past into the culture of the "New China" are also responsible for the considerable attention being given to Ssu-ma Ch'ien and his Shih-Chi (6.4.8 and 6.4.9).

While historical articles are published in a great number of journals, 6.4.15-18 are the most important historical periodicals in the People's Republic of China. Among them Li-shih yen-chiu occupies the primary position. As of the time of this writing, these four journals and their counterparts in many other disciplines have not been generally available outside of China since the end of 1959, when their exportation from the People's Republic of China was stopped ostensibly on the grounds that a shortage of paper required a reduction in the number of copies being printed and thus available for export.

6.4.1 Shanghai Branch of the Chinese Historical Association (Chung-kuo shih-hsueh hui 中国史学会) ed., Tsen-yang hsueh-hsi tsu-kuo ti li-shih 怎樣学習祖国的历史 (How to study the history of our fatherland; Shanghai: Hua-tung jen-min ch'u-pan-she, 1953), 86 pp.

Six articles previously published in periodicals (Hsin chien-she 新建設 , Hsueh-hsi 学習 , and Chung-kuo ch'ing-nien 中国青年) have been revised for this collection. Jung Meng-yuan writes on "Patriotism and historical science" and on "Study Stalin's Dialectical and Historical Materialism." The other pieces are: Chien Po-tsan 翦伯贊 , "How to study Chinese history"; Chin Cho-jan 金燦然 , "Love the history of our fatherland"; Yeh Hu-sheng 葉蠖生 , "The basic viewpoint for the study of Chinese modern history"; and Hu Hua 胡华 , "How to study the history of China's new-democratic revolution." While the purpose of this collection is obviously to persuade the reader to follow the historical doctrines of Marx, Lenin, Stalin, and Mao, the arguments of the several authors are not equally doctrinaire. Jung, for example, criticizes the indiscriminate application of certain Marxist concepts among his "comrades," while Hu in his exposition on the leadership of the CCP and Mao maintains an unwavering exhortative tone. Jung was later to be denounced as a "rightist." The book is a guide to the points of emphasis in mainland Chinese historiography. (80,000)

6.4.2 Lü Chen-yü 吕振羽, Shih-hsueh yen-chiu lun-wen chi 史学研究論文集 (Collected essays on historical research; Shanghai: Hua-tung jen-min ch'u-pan-she, 1954), 3 + 192 pp.

These six articles by a veteran Marxist historian are entitled: (1) "My fragmentary views concerning historiographical methodology"; (2) "On social consciousness and the great cultural heritage of our great fatherland"; (3) "Answers to questions such as 'How should we study Chinese history?'"; (4) "Several questions concerning the development of Chinese society"; (5) "On the Taiping revolutionary movement"; and (6) "Concerning the economic principles of Soviet socialism." They were originally reports or lectures delivered before local political study classes in Manchuria between 1949 and 1953, and are replete with Marxist-Leninist-Stalinist-Maoist clichés. (32,000)

6.4.3 Chao Li-sheng 趙儷生, Shih-hsueh hsin t'an 史学新探 (A new inquiry into historiography; Shanghai: Hsin-chih-shih ch'u-pan-she, 1954), 111 pp. (15,000)

6.4.4 Chin Yü-fu 金毓黻, Chung-kuo shih-hsueh shih 中国史学史 (A history of Chinese historiography; Shanghai: Commercial Press, 1957), 8 + 273 pp.

Originally published in Chungking in 1944, this revised edition ends with the Ch'ing dynasty and omits a forty-six-page chapter on the modern period which appeared in the first edition. The author apologizes for not having established a "dialectical-materialistic viewpoint" back in 1938 when he first completed the manuscript of the book and for his consequent failure to indicate the "mainstream of development" in Chinese historiography. Penind the production of a work on the "new model," this reprint is offered as a temporary reference. (8,000)

6.4.5 Cheng Ho-sheng 鄭鶴聲, Chung-kuo shih pu mu-lu hsueh 中国史部目録学 (The classification of Chinese historical works; Shanghai: Commercial Press, 1956), 5 + 265 pp.

A revised edition of a 1933 publication, this volume retains substantially the original descriptions of traditional Chinese historical genres and works and their classification. However, the original chapter on "modern cataloguing" has been drastically revised on the basis of more recent cataloguing systems. Deleted from this edition are a chapter on "historical forms" and the original conclusion (which contained some "erroneous" theories on book classification derived from English and American sources). The new concluding chapter, "New tendencies in the classification of books on historical science," takes into account such guiding principles of classification as "political ideology." (3,300)

6.4.6 Nanking Office for the Organization of Historical Materials, Third Office of the Institute of Historical Research, Chinese Academy of Sciences, Li-shih tang-an ti cheng-li fa 歷史檔案的整理法 (The method of organizing historical archives; Peking: Jen-min ch'u-pan-she, 1957), 4 + 162 pp.

This book contains practical details related to the organization of historical archives based on the experience of a group of archivists at the Institute of Historical Research, who decry pre-Communist archival work in China and claim to be indebted to the Soviet example. They discuss the significance of planning, procedures, and the desired results of archival management. While heavily ideological, this volume contains some incidental data on archives of the late Ch'ing and Republican periods. (3,500)

6.4.7 Hao Chien-liang 郝建樑 and Pan Shu-ko 班書閣, eds., Chung-kuo li-shih yao-chi chieh-shao chi hsuan-tu 中国历史要籍介紹及選讀 (An introduction to and selected readings in major Chinese historical works; Peking: Kao-teng chiao-yü ch'u-pan-she, 1957), 217 pp. (15,000)

6.4.8 Chi Chen-huai 季鎮淮, Ssu-ma Ch'ien 司馬遷 (Shanghai: Shang-hai jen-min ch'u-pan-she, 1955), 2 + 140 pp.

A semi-popular biography. (30,000)

6.4.9 Editorial Committee of the journal Wen shih che 文史哲, ed., Ssu-ma Ch'ien yü Shih-chi 司馬遷与史記 (Ssu-ma Ch'ien and the Shih-chi; Peking: Chung-hua shu-chü, 1957), 188 pp. (12,100)

6.4.10 Wang Ming-sheng 王鳴盛, Shih-ch'i shih shang-chüeh 十七史商榷 (A critical study of seventeen dynastic histories; Shanghai: Commercial Press, 1959), 2 vols., 84 + 1158 pp.

This is a reissue of the 1937 Commercial Press edition of the major scholarly work of Wang Ming-sheng (1722-98), first printed in 1787. The 100 chüan deal with the histories beginning with Shih chi and ending with Wu-tai shih 五代史. In his preface, Wang states that he has compared the various editions of these histories in an attempt to correct errors, supply missing parts, weed out corruptions, explain ambiguities, weigh contradictions, and, in so doing, make objective historical facts available to the reader. (1,900)

6.4.11 Chao I 趙翼, Nien-erh shih cha-chi 廿二史劄記 (Notes on twenty-two histories; Peking: Commercial Press, 1958), 17 + 782 + 14 pp.

A reprint, with no changes, of a 1937 punctuated edition of this classic of Ch'ing historical scholarship. It contains topical studies on problems in actually twenty-four histories, including *Ming shih* 明史 (Ming history). Chao I lived from 1727 to 1814. (1,500)

6.4.12 Chao I 趙翼, *Kai-yü ts'ung k'ao* 陔餘叢考 (Notes collected while visiting parents; Shanghai: Commercial Press, 1957), 20 + 979 pp.

This is a punctuated reprint (based on a 1790 edition) of a work consisting of notes on a wide variety of historical and related subjects by an important Ch'ing scholar (1727-1814). Of the forty-three *chüan*, *chüan* 5-15, for example, are on the topic of historiography; 16-21 on historical records; 25, on reign periods; 26-27, on official systems; and 28-29, on the examination system. (4,000)

6.4.13 Ch'ien Ta-hsin 錢大昕, *Nien-erh shih k'ao i* 廿二史考異 (Collations of twenty-two standard histories; Peking: Commercial Press, 1958), 2 vols., 3 + 1608 pp.

A reprint, with no changes, of a 1937 punctuated edition of this important Ch'ing study of textual problems in the standard histories. It is in the same format as Chao I's *Nien-erh shih cha-chi* (6.4.11). The two T'ang histories are counted as one and *Ming shih* is not included, making twenty-two. Ch'ien Ta-hsin lived from 1728 to 1804. (1,400)

6.4.14 Chang Hsueh-ch'eng 章學誠, *Wen shih t'ung-i* 文史通義 (General principles of literature and history; Peking: Ku-chi ch'u-pan-she, 1956), 13 + 370 pp.

Chang Hsueh-ch'eng, *Chiao-ch'ou t'ung-i* 校讎通義 (General principles of bibliography; Peking: Ku-chi ch'u-pan-she, 1956), 9 + 94 pp.

The two most important extant works of the Ch'ing historian Chang Hsueh-ch'eng (1738-1801) are reprinted, in two separate volumes, based on a 1921 edition (Chang-shih i shu 章氏遺書), but containing some additional essays not included in that edition. The publisher's note is the same for both volumes, praising Chang's contributions to historiography and describing him as a "materialist." (4,000; 3,500)

6.4.15 Li-shih yen-chiu 历史研究 (Historical research; Peking: K'o-hsueh ch'u-pan-she, first number, Jan. 1954), bi-monthly, 1954-55; monthly since Jan. 1956.

The leading historical journal in the People's Republic of China. Table of contents in Russian and English as well as in Chinese.

6.4.16 Li-shih chiao-hsueh 历史教学 (The teaching of history; Tientsin: Li-shih chiao-hsueh yueh-k'an-she, first number, Jan. 1951), monthly.

A publication for middle school teachers of history containing semi-popular articles by leading scholars on the major problems currently of interest to mainland historians.

6.4.17 Chin-tai shih tzu-o-liao 近代史資料 (Source materials on modern [Chinese] history; Peking: K'o-hsueh ch'u-pan-she, first number, Aug. 1954), bi-monthly.

The contents consisted of newly published documents and other source materials on the history of modern China. In 1959, this publication ceased to be a journal and was transformed into a series of volumes of source materials on single subjects, issued from time to time.

6.4.18 Wen shih che 文史哲 (Literature, history, and philosophy; Tsingtao: Shantung University, first number, May[?] 1951), monthly.

Contains scholarly articles on a wide range of historical subjects. Not generally available in the United States.

LIST OF PUBLISHERS

Chiang-su jen-min ch'u-pan-she 江苏人民出版社
Ch'ing-kung-yeh ch'u-pan-she 輕工业
Ch'ing-nien ch'u-pan-she 青年
Chung-hua shu-chü 中华書局
Chung-kuo ch'ing-nien ch'u-pan-she 中国青年
Chung-kuo jen-min ta-hsueh ch'u-pan-she 中国人民大学
Chung-nan jen-min ch'u-pan-she 中南人民
Chung-wai ch'u-pan-she 中外
Ch'ung-ch'ing jen-min ch'u-pan-she 重庆人民
Ch'ün-lien ch'u-pan-she 群联
Commercial Press (Shang-wu yin-shu kuan) 商務印書館

Fa-lü ch'u-pan-she 法律
Fu-chien jen-min ch'u-pan-she 福建人民

Hai-yen shu-tien 海燕書店
Hsin chih-shih ch'u-pan-she 新知識
Hsin-chih shu-tien 新知
Hsin Chung-kuo shu-chü 新中国
Hsin-hua ch'u-pan-she (Hong Kong) 新華
Hsin-hua shu-chü 新華
Hsin-hua shu-tien 新華
Hsin min-chu ch'u-pan-she 新民主
Hsin wen-i ch'u-pan-she 新文艺
Hsin-ya shu-tien 新亞
Hsueh-hsi sheng-huo ch'u-pan-she 学習生活
Hsueh-hsi tsa-chih she 学習杂誌社
Hua-nan jen-min ch'u-pan-she 华南人民
Hu-nan t'ung-su tu-wu ch'u-pan-she 湖南通俗讀物
Hu-pei jen-min ch'u-pan-she 湖北人民

Hua-tung jen-min ch'u-pan-she 华東人民

I-ch'ang shu-chü 益昌

Jen-min chiao-t'ung ch'u-pan-she 人民交通
Jen-min chiao-yü ch'u-pan-she 人民教育
Jen-min ch'u-pan-she 人民
Jen-min mei-shu ch'u-pan-she 人民美術
Jen-min wen-hsueh ch'u-pan-she 人民文学

K'ai-ming shu-tien 開明
Kao-teng chiao-yü ch'u-pan-she 高等教育
K'o-hsueh ch'u-pan-she 科学
Ku-chi ch'u-pan-she 古籍
Ku-tien wen-hsueh ch'u-pan-she 古典文学
Kuang-tung jen-min ch'u-pan-she 廣東人民
Kung-jen ch'u-pan-she 工人

Lao-tung ch'u-pan-she 勞動
Li-shih chiao-hsueh yüeh-k'an she 历史教学月刊社
Lung-men lien-ho shu-chü 龙门联合

Min-tsu ch'u-pan-she 民族

Pei-ching ch'u-pan-she 北京

San-lien shu-tien (full name: Sheng-huo tu-shu hsin-chih san-lien shu-tien) 生活讀書新知三联
Shan-tung jen-min ch'u-pan-she 山東人民
Shang-hai chiao-yü ch'u-pan-she 上海教育
Shang-hai ch'u-pan kung-ssu 上海出版公司
Shang-hai jen-min ch'u-pan-she 上海人民
Shang-tsa ch'u-pan-she 上雜
Shen-chou kuo-kuang she 神州国光社
Shensi jen-min ch'u-pan-she 陕西人民

Shih-chieh chih-shih she 世界知識社
Shih-hsi ch'u-pan-she 实習
Shih-tai shu-chü 时代
Ssu-lien ch'u-pan-she 四联

T'ai-lien ch'u-pan-she 泰联
T'ang-ti ch'u-pan-she 棠棣
Ti-t'u ch'u-pan-she 地圖
T'ien-chin jen-min ch'u-pan-she 天津人民
T'ien-chin t'ung-su ch'u-pan-she 天津通俗
Tso-chia ch'u-pan-she 作家
T'ung-chi ch'u-pan-she 統計
T'ung-su tu-wu ch'u-pan-she 通俗讀物

Wen-tsung ch'u-pan-she (Hong Kong) 文宗
Wen-tzu kai-ko ch'u-pan-she 文字改革
Wen-wu ch'u-pan-she 文物
Wu-shih nien-tai ch'u-pan-she 五十年代

INDEX

The reader should also consult the section headings in the table of contents for general topics, which are not duplicated here.